Lecture Notes in Networks and Systems

Volume 432

The series "Lecture Notes in Networks and Systems" publishes the latest developments in Networks and Systems—quickly, informally and with high quality. Original research reported in proceedings and post-proceedings represents the core of LNNS.

Volumes published in LNNS embrace all aspects and subfields of, as well as new challenges in, Networks and Systems.

The series contains proceedings and edited volumes in systems and networks, spanning the areas of Cyber-Physical Systems, Autonomous Systems, Sensor Networks, Control Systems, Energy Systems, Automotive Systems, Biological Systems, Vehicular Networking and Connected Vehicles, Aerospace Systems, Automation, Manufacturing, Smart Grids, Nonlinear Systems, Power Systems, Robotics, Social Systems, Economic Systems and other. Of particular value to both the contributors and the readership are the short publication timeframe and the world-wide distribution and exposure which enable both a wide and rapid dissemination of research output.

The series covers the theory, applications, and perspectives on the state of the art and future developments relevant to systems and networks, decision making, control, complex processes and related areas, as embedded in the fields of interdisciplinary and applied sciences, engineering, computer science, physics, economics, social, and life sciences, as well as the paradigms and methodologies behind them.

Indexed by SCOPUS, INSPEC, WTI Frankfurt eG, zbMATH, SCImago.

All books published in the series are submitted for consideration in Web of Science.

For proposals from Asia please contact Aninda Bose (aninda.bose@springer.com).

More information about this series at https://link.springer.com/bookseries/15179

Arthur Gibadullin
Editor

Digital and Information Technologies in Economics and Management

Proceedings of the International Scientific and Practical Conference "Digital and Information Technologies in Economics and Management" (DITEM2021)

 Springer

Editor
Arthur Gibadullin
Institute of Industry Management
State University of Management
Moscow, Russia

ISSN 2367-3370 ISSN 2367-3389 (electronic)
Lecture Notes in Networks and Systems
ISBN 978-3-030-97729-0 ISBN 978-3-030-97730-6 (eBook)
https://doi.org/10.1007/978-3-030-97730-6

This Springer imprint is published by the registered company Springer Nature Switzerland AG
The registered company address is: Gewerbestrasse 11, 6330 Cham, Switzerland

Preface

The conference was held with the aim of summarizing international experience in the field of digital development of the economy and management, the introduction of information technologies and systems in the organizational processes of managing corporations and individual industries.

The conference presents scientific research aimed at solving a set of problems in the field of information technologies in economics and management, namely

- Development of scientific and practical potential in order to formulate proposals for the introduction of digital and information technologies in the economy and management;
- Popularization of fundamental and applied research in the field of information technologies and systems;
- Formation of recommendations aimed at improving computer models, information technology, engineering, innovative and digital technologies in economics and management.

The International Scientific and Practical Conference "Digital and Information Technologies in Economics and Management" (DITEM2021) was held on November 2, 2021, on the Microsoft Teams platform due to COVID-19.

The conference addressed issues of information, digital and intellectual technologies in economics and management. A distinctive feature of the conference is that it presented reports of authors from Italy, South Korea, Poland, Armenia, Republic of Belarus and the Russian Federation. Researchers from different countries presented the process of transition of economic activities to the information and digital path of development and presented the main directions and developments that can improve the efficiency and development of the economy and management.

The sections were moderated by:

- Gibadullin Artur Arturovich—Candidate of Sciences, Associate Professor of the Department of Economics and Management in the Fuel and Energy Complex of the State University Management, Moscow, Russia;

- Morkovkin Dmitry Evgenievich—Candidate of Sciences, Associate Professor of the Financial University under the Government of the Russian Federation, Moscow, Russia.

The organizing committee decided not to postpone the conference, as COVID-19 is gaining momentum, in connection with which the conference took place at the appointed time.

Thus, the conference made it possible to formulate scientific recommendations on the use of information, computer, digital and intellectual technologies in industry and fields of activity that can be useful to state and regional authorities, international and supranational organizations and the scientific and professional community.

Acknowledgments. The organizing committee of the conference expresses its gratitude to the staff of Springer Nature, who helps to place the proceedings of scientific conferences in international journals and provide support for the publication of materials. In addition, the organizing committee would like to thank the conference participants, reviewers and everyone who helped organize this conference and form the presented volume for publication in Springer Nature.

Peer-Review Declaration

All papers published in this volume have been peer-reviewed through processes administered by the editors. Reviews were conducted by expert referees to the professional and scientific standards expected of a proceedings journal published by Springer Nature.

- Type of peer review: Double-blind
- Describe criteria used by reviewers when accepting/declining papers. Was there the opportunity to resubmit articles after revisions?

All submitted articles underwent double anonymous review. The reviewers were based on the assessment of the topic of the submitted materials, the relevance of the study, the scientific significance and novelty, the quality of the materials and the originality of the work. Authors could revise the article and submit it again for review. Reviewers, program committee members and organizing committee members did not enter into discussions with the authors of the articles.

- Conference submission management system:

  1. Sending materials to the organizers of the conference (ditem2021@mail.ru, managed by Gibadullin Arthur);
  2. Verification of compliance with the subject of the conference;
  3. Checking materials for plagiarism using two systems;
  4. Submitting for double-blind review;
  5. Checking the quality of translation of the article into English;

6. Checking for the correct design of materials;
7. Submission of materials to the publisher.

- Number of submissions received: 67
- Number of submissions sent for review: 67
- Number of submissions accepted: 23
- Acceptance rate (number of submissions accepted/number of submissions received X 100): 34,3%
- Average number of reviews per paper: 2
- Total number of reviewers involved: 21
- Any additional info on review process: To be accepted for publication, the material must have the following requirements: compliance with the conference theme; a high level of originality of the text (at least 80%); positive expertise of reviewers; good level of English and well-formed material.
- Contact person for queries: Gibadullin Arthur—Candidate of Sciences, State University of Management, Moscow, Russian Federation—Executive Secretary of the Organizing Committee (ditem2021@mail.ru).

Arthur Gibadullin

Contents

Economic Modeling of the Impact of Human Capital on Economic Growth and Competitiveness

Felix Ereshko[1]([✉]), Ivan Kamenev[1], and Alexey Shimansky[2]

[1] Computing Center of the Russian Academy of Sciences, Federal Research Center "Informatics and Control" RAS, 40, Vavilova Street, Moscow 119991, Russia
fereshko@yandex.ru
[2] Peoples' Friendship University of Russia (RUDN University), 6, Miklukho-Maklaya Street, Moscow 117198, Russian Federation

Abstract. The article discusses the problem of constructing a dynamic macroeconomic general equilibrium model designed to explain the relationship between investment in material and human capital with economic growth for the Russian economy. An overview of approaches to modeling human capital is given and the feasibility of developing a new approach to its modeling within the framework of general equilibrium models is substantiated. The general structure of the model is given, the problem of the household is posed and solved, general approaches to and identification of the model, the study of the problem of the firm, taking into account state policy, are developed. The main feature of the proposed model is the possibility of considering in it the choice between investments in material and human capital from the point of view of the scale of the impact on economic growth. An essential minor feature is that the intertemporal choice of a household is modeled without using the financial market, which can be considered an advantage when modeling the Russian economy, in whose financial market households are poorly represented.

Keywords: Human capital · Mathematical modeling · Investments · Education market · Health care market · Labor market · General equilibrium models · DSGE models · DSOER · Economic growth

1 Introduction

Subsequent paragraphs, however, are indented. Starting with the classical works of G. Becker and T. Schultz, the concept of human capital assumes the application of the same approaches to the factors of production, which are carried by a person, as to the traditionally considered material capital. The main scientific ideas of these studies are: the interpretation of human capital as an asset that brings income to the owner, and the corresponding modeling of the investment scheme; expanding the concept of capital to the social structure of society and justifying the trust resulting from it; transition to lifetime indicators of return on (human) capital.

© The Author(s), under exclusive license to Springer Nature Switzerland AG 2022
A. Gibadullin (Ed.): DITEM 2021, LNNS 432, pp. 1–14, 2022.
https://doi.org/10.1007/978-3-030-97730-6_1

Microeconomic modeling of human capital is a fairly developed area, in general following the approach laid down in the classical model by J. Mintzer [1, 2], according to which human capital brings an individual additional (labor) income compared to the income provided by work experience. On the basis of this idea, a whole class of models has developed, describing the choice of the amount of investment by parents in the human capital of their children. These models (both classical, see [3–7] and modern, see, for example, [8–13]) substantiate the economic feasibility of the existence of education and health care markets and their state funding. The main discussion in this direction of the theory of human capital is centered around the question of to what extent the return on human capital is explained by the increase in employee productivity, and not by signaling and discrimination in the labor market.

The second approach to modeling human capital is economic-theoretical. The authors working in this direction do not aim to reproduce the dynamics of a particular national economy. Instead, they propose dynamic models with a high level of abstraction that reveal the possibility of using human capital to explain non-linear patterns in economic development.

At the macroeconomic level, another class of models has emerged that substantiate government policy in the field of human capital. These researchers consider the contribution of human capital to economic development. To do this, they form various indices of human capital or its components for a country or region, and compare the level and dynamics of economic development with the development of human capital. Models of this type have made a significant contribution to the fact that developed countries are no longer considering social policy solely as a direction of spending, but also as a form of investment in economic development.

The fundamental limitation of such models is that they are not choice models. The development of human capital (both through private and public funding) requires resources that could be invested in the development of material capital, and that models of this class are either not considered or are evaluated purely statistically. This makes the existing macroeconomic models of limited applicability for studying the problems of economic growth, and determines the relevance of developing a new class of models in which human capital as a factor of production would be directly included in the model of economic growth.

2   Modeling Economic Growth and the Impact of Human Capital on it

There are two main approaches to modeling economic growth in macroeconomics. The first one is purely macroeconomic models (for example, the Keynesian-neoclassical synthesis model ISLM + ADAS and its analogues), in which the state acts as the only subject of decision-making, while economic activity itself is reduced to a system of balances that describe different economic processes. The quality of such models is checked econometrically, according to the results of their calibration against the observed macroeconomic indicators. Economic growth is operationalized in them as the dynamics of potential GDP, i.e. the maximum possible value of GDP, achieved with a "correct" economic policy. As a rule, it is modeled through the demographic pyramid (or, to put it simply, the net reproduction rate), the capital investment function, and a constant rate of technological progress. Thus, in these models, economic growth is a function of investment in capital, and the government's task to maintain economic growth is reduced to creating conditions for maximizing these investments.

The second approach is general equilibrium models, which imply a combination of the macroeconomic level of analysis with the tasks of agents, although the agents are described at the highest possible level of aggregation. Such models are more difficult to construct and research, but they allow us to consider the economy in a more comprehensive way, explaining the logic of the choice not only of the state, but also of other economic entities. Models of this type for the Russian economy are being developed by a number of Russian economists including the Bank of Russia.

The general features of general equilibrium models are as follows. There are three agents in the model: households, firms, and states. These agents interact in several markets (most often, labor markets, financial capital and goods). Each agent solves its own optimization problem within the framework of a certain expectation model. The capabilities of an economy are described by a production function, which describes the transformation of resources into products.

In the simplest version of general equilibrium models, the product is one and the resources are exogenously given. However, this version is unacceptable neither for modeling economic growth, nor for modeling human capital. Let us list the requirements for the general equilibrium model that make it suitable for solving the problem under consideration:

1. The model must contain dynamic functions that describe the change in both resources over time (i.e., endogenous economic growth). Moreover, in the existing general equilibrium models, the dynamic function of investment in (material) capital is often used, while labor is modeled exclusively by the demographic function, which does not allow studying the influence of human capital on the national economy. A side effect of this approach is the increased attention of the political leaders of many countries (including Russia) to the problem of stimulating the birth rate (maternity capital) and expanding the boundaries of the working age (pension reform), in comparison with the problems of education and healthcare: these areas are simply not taken into account in the models, according to who make macroeconomic decisions. Accordingly, the main research question is how the dynamic function of human capital can be modeled in the general equilibrium model.

2. In order to include human capital in the general equilibrium model, the model must contain 2 products. This is due to the fact that material and human capital certainly cannot be transformed into each other, i.e. existing capital cannot be redistributed between "material" and "human" forms. This is substantively explained by the fact that only a person can be the bearer of human capital, and it can be formed exclusively with the help of services. From the point of view of model dynamics, this means that material and human capital must be designated by different variables and described by independent dynamic equations.

3. The model should include markets for human capital (services) and material goods, which describe not only the volume of current consumption, but also the volume of investment in human and material capital.

4. The model should provide for the possibility of calibrating the coefficients associated with human capital, according to macroeconomic data, just as it is done in the existing general equilibrium models for physical capital.

Note that today we do not know of general equilibrium models that meet these four requirements simultaneously (purely macroeconomic models, by definition, cannot match them).

3   Operationalization of Human Capital

Human capital is modeled as a person's ability to create a useful product. It depends on the health and education of the person. Health determines the amount of time a person can devote to work, and the level of education - qualifications, the ability to perform a larger amount of work per unit of time.

The human capital of the model can be identified by the United Nations Human Development Index (HDI), which provides a generalized description of the country's health and education. However, in addition to education and health, the HDI has a third component (income index) that must be removed from the index before use.

$$K_n^l = LEI_n^{1/2} EI_n^{1/2} \qquad (1)$$

K_n^l – human capital;
LEI_n – Life Expectancy Index, calculated by the United Nations for Russia in the year n;
EI_n – Education Index, calculated by the United Nations for Russia in the year n.

Identification by HDR is deliberately rude due to the known shortcomings of this index (in particular, it does not take into account the qualifications acquired in the process of work), but it gives the most stable series of data on human capital.

With regard to Russia, the values of these indices are as follows (Table 1):

4   The Static Part of the General Equilibrium Model with Human Capital

The model contains three macroeconomic agents: households, firms, and the state. Each agent, except for the state, solves the problem of intertemporal choice taking into account the discount factor z from the moment of making the decision n_0 for a period of time τ (planning horizon)

$$z_n = \left(\frac{1}{1+r}\right)^{\Delta n} = \left(\frac{1}{1+r}\right)^{n-n_0}, \Delta n \le \tau \qquad (2)$$

The intertemporal choice of the household consists in the purchase of consumer goods C_n or human capital development services E_n taking into account the living

Table 1. Model identification data: Human capital (K_n^l) for n ∈ [1990, 2018].

n	1990	1991	1992	1993	1994	1995	1996	1997	1998	1999
EI	0.734	0.729	0.719	0.711	0.702	0.701	0.702	0.704	0.703	0.710
LEI	0.739	0.732	0.724	0.717	0.712	0.708	0.705	0.702	0.699	0.696
Kl	0.736	0.730	0.721	0.714	0.707	0.704	0.703	0.703	0.701	0.703
n	2000	2001	2002	2003	2004	2005	2006	2007	2008	2009
EI	0.721	0.727	0.733	0.740	0.746	0.752	0.759	0.767	0.774	0.771
LEI	0.694	0.692	0.692	0.694	0.698	0.704	0.711	0.720	0.730	0.740
Kl	0.707	0.709	0.712	0.716	0.721	0.727	0.734	0.743	0.751	0.755
n	2010	2011	2012	2013	2014	2015	2016	2017	2018	2019
EI	0.780	0.789	0.797	0.803	0.807	0.813	0.817	0.822	0.824	–
LEI	0.749	0.759	0.768	0.777	0.785	0.792	0.797	0.802	0.806	–
Kl	0.764	0.773	0.782	0.789	0.795	0.802	0.806	0.811	0.815	–

wage C_0 and labor depreciation C_e. The criterion for the optimal choice is the utility of consumption on the planning horizon (3) with a constraint on the household budget (4):

$$U_n^c = \sum_{n=n_0}^{\tau} C_n z_n \rightarrow max \tag{3}$$

$$C_n p_n^c L_n + E_n p_n^e L_n + C_e p_n^c e_n^2 L_n + C_0 p_n^c L_n = w_n (1 - t_l) K_n^l e_n L_n \tag{4}$$

here K_n^l – human capital (altered by an endogenous process), and L_n – number of labor resources, people (changes by an exogenous process, demographic), e_n – the share of the productively employed in the number of labor resources established by households. In what follows, referring to the level of employment, we mean exactly e_n. It should be noted that the form of the budgetary constraint involves a transition to per capita calculation and a reduction in L_n. Fractional values L_n interpreted as partial work capacity of the individual.

Households choose between consuming now and investing in human capital that will enable them to earn more in the future. For the household, prices for material goods p_n^c and services for the development of human capital p_n^e and labor taxes t_l are exogenous constants (that is, households do not have information for their meaningful forecasting).

Firms tackle the problem of intertemporal choice between dividends D_n and investment I_n, maximizing dividends on the planning horizon going to the prestigious consumption of the owners of the firm (5), under the constraints of the production function with production factors by human capital K_n^l, physical capital K_n^m and busy labor resources $e_n(w)L_n$ (6) and budget function (7):

$$U_n^b = \sum_{n=n_0}^{\tau} D_n * z_n \rightarrow max \tag{5}$$

$$D_n(1 - t_d) + I_n p_n^c = Y_n^c(p_n^c)p_n^c + Y_n^e(p_n^e)p_n^e$$
$$-e_n(w_n)w_n K_n^l L_n - (Y_n^e + Y_n^c)(1 - t_Y) \tag{6}$$

$$Y_n^e + Y_n^c = a\left(K_n^l e_n(w_n)L_n\right)^\alpha * \left(K_n^m Z_n\right)^{1-\alpha} \tag{7}$$

Since the Russian economy is modeled with a high level of income inequality and capital concentration, the profits of firms are not returned to households. The model assumes that consumed prestigious goods require a minimum consumption of factors of production (their market value is many times higher than the cost of production), therefore, prestigious consumption is not included in the balance of material goods. Due to their small number, the owners also do not influence the dynamics of human capital.

In addition to tax rates on dividends t_d and on production t_Y, an exogenous shock Z_n is also introduced here, reflecting changes in the external economic environment. This is due to the fact that the production of material goods in Russia is strongly associated with foreign economic activity (the intermediate product of the raw materials industries is exported, the intermediate product of the industry is imported).

The general meaning of the manufacturer's problem is that he chooses between the current income D_n and the future income that grows at the expense of I_n. A large number of controls (w_n, Y_n^e, Y_n^c, I_n), connected with each other, call into question the solvability of the manufacturer's problem, but it is computable.

The state determines macroeconomic policy with the following constraints (budget):

$$D_n t_d + (Y_n^e + Y_n^c)t_Y + w_n t_l K_n^l e_n L_n = E_n^g p_n^e L_n + C_g p_n^c + I_g p_n^c \tag{8}$$

Here C_g there is government consumption (maintenance of the state apparatus and other autonomous expenses, i.e. expenses that do not affect the capacity of the economy). On the contrary, government investments in material and human capital can contribute to economic growth. However, a crowding-out effect is also possible in the model: for example, if households consider government investments in human capital sufficient, then they will not make their own investments. In the model, firms and households do not have the information necessary for meaningful forecasting of public policy, and they consider it constant.

In accordance with the number of goods and factors of production, the model includes three markets, one of which is set implicitly (by the rules for choosing a firm and a household), and two are described by balance equations.

Labor market (implicitly defined) ensures equality e_n, established by households, and e_n, used by firms.

The human capital (services) market is set by the balance equation:

$$E_n(p_n^c, p_n^e, w_n)L_n + E_n^g L_n = Y_n^e(p_n^c, p_n^e, w_n) \tag{9}$$

The market of goods is given by the balance equation:

$$C_n(p_n^c, p_n^e, w_n)L_n + C_e(p_n^c, p_n^e, w_n)e_n^2 * L_n + C_0 L_n + C_g$$
$$+ I_n(p_n^c, p_n^e, w_n) + I_g = Y_n^c(p_n^c, p_n^e, w_n) \tag{10}$$

The presented system of balances is investigated for each n. The task of the individual is solvable analytically, which allows the use of individual controls $C_n(p_n^c, p_n^e, w_n)$, $E_n(p_n^c, p_n^e, w_n)$, $e_n(p_n^c, p_n^e, w_n)$, as functions of prices determined in the firm's problem. The study of the company's problem, in turn, is possible only by methods of simulation, which requires a separate publication. It is supposed to use for this the Metric Analysis Method in the version of the Identification Sets Method: calculate the firm's functional for different values of the wage rate, firm investments and the distribution of output between material goods and services for the development of human capital. This will allow calculating with high accuracy the optimal value of the firm's controls under various regimes of state policy (including those close to Russian conditions).

$$K_{n+1}^m = K_n^m + f_m(I_n + I_g) - A_m K_n^m \tag{11}$$

Dynamics of human capacities of the national economy (human capital):

$$K_{n+1}^l = K_n^l + f_l \frac{(E_n + E_n^g)}{L_n} - A_l K_n^l \tag{12}$$

Here f_m и f_l – capacity ratios of human and physical capital (the amount of investment required to form a unit of capital). A_m and A_l – depreciation rates (depreciation and obsolescence) of human and physical capital. All of them are entered as exogenous (calibration) constants and must be determined during model identification.

An important advantage of general equilibrium models is the presence in them of endogenous economic growth (in this case, explained by the development of material and human capital). In the model under consideration, it is defined as the change in potential total output: $LRY = a(K_n^l L_n(w_n))^\alpha (K_n^m Z_n)^{1-\alpha}$.

Note that it coincides with $Y_n^e + Y_n^c$ under the condition of full employment, and thus we can calculate for a specific year n even without solving the full problem of the firm. However, the complete dynamics of the model naturally requires finding an approximate solution to the firm. Increase as K_n^l, and K_n^m makes the economy more competitive, but from the point of view of the state, the main interest is the proportion between investments in material and human capital, which ensures the highest growth rates.

It is this opportunity that is the main advantage of the model: unlike all models of human capital proper, this model does not represent investment in it as an a priori positive phenomenon, because investing in human capital, one has to sacrifice the development of material capital (and vice versa). On the other hand, if, for any reason, firms or households are not ready to invest in their capital, the state can "do it for them".

5 Household Behavior and Investment in Human Capital

Within the framework of the model, a household makes two decisions: how much to work, and how to distribute the earned between the acquisition of consumer goods and services for the development of human capital. Since the household uses simple nominal expectations, prices for it are exogenous constants (13).

$$C_n p^c + E_n p^e = e_n w(1 - t_l) K_n^l - C_e p^c e_n^2 - C_0 p^c \tag{13}$$

This formulation allows us to split the household task into two: the task of the employee household and the task of intertemporal choice. These tasks are independent: in any case, a household needs to maximize its disposable income (minus depreciation of labor, i.e. additional expenditure of energy and health in the labor process). This allows us to set an optimization problem for a function of one variable (14).

$$F(e_n) = e_n w(1 - t_l) K_n^l - C_e p^c e_n^2 - C_0 p^c \to \max \qquad (14)$$

$$F(e_n)' = w(1 - t_l) K_n^l - 2 C_e p^c e_n = 0 \qquad (15)$$

To simplify the subsequent notation, we denote the found optimal employment level as e^e.

$$e^e(w, Kl_n) = \begin{cases} \frac{w(1-t_l)K_{nn}^l}{2C_e p^c}, \frac{w(1-t_l)K_n^l}{2C_e p^c} \le 1 \\ 1, \frac{w(1-t_l)K_n^l}{2C_e p^c} > 1 \end{cases} \qquad (16)$$

It is essential that at low values of human capital, the choice of households is not stationary; the employment rate will grow with the growth of human capital, because pays for more and more depreciation. At large values of human capital, it ceases to affect employment: the limiting factor is not the amortization of human capital, but the size of the population. In what follows, however, we will consider the mode of the model in which the control e^e stationary. This is due to mathematical, statistical and substantive considerations. Mathematically, a permanent control decision e_e allows the household to analytically solve the intertemporal choice problem, while with a time variable e_e the analytical solution is difficult. The problem can be easily investigated through simulation experiments, but this way of research is unrealistic for households. Unlike firms, for which failed experiments and bankruptcy are the norm, households cannot afford sudden death.

Substantially, the micro-descriptions of the labor supply, on which the general equilibrium models are based (including in this model), proceed from the negative utility of labor (the "laziness" of individuals). This is due to the need to amortize labor (a person gets tired and must recuperate after labor), but no less important is the stable emotional connection between labor and fatigue developed in households. Accordingly, the ability of households to predict an increase in the level of their labor activity is questionable (it contradicts the psychology of a "lazy" worker). At the macroeconomic level, this argument is insufficient, since an increase in the intensity of labor use is possible not only due to overtime work of employed people, but also due to previously unemployed and economically inactive. However, sociological studies in Russia show that households predominantly expect and fear job loss, and these expectations persist both during periods of crises and boom.

Finally, the statistical argument for using a permanent management solution e_e will differ for the Russian economy and for developed countries. In developed countries with persistent, relatively low unemployment, the number of unemployed and economically inactive people expecting that they may enter the labor market is inevitably lower than the number of employed people expecting a possible reduction in their employment.

This mathematical proportion changes for a short time during an economic crisis, but during a crisis the expectations of both employed and unemployed are more pessimistic. The phase of recovery growth is noticed (according to statistics, news background, etc.) with a delay of six months or a year, when employment is restored to normal levels. On the other hand, in the Russian economy, the discussed fluctuations in employment during the crisis period are deliberately insignificant, the Russian labor market is characterized by the rigidity of employment and price flexibility. This feature becomes an additional argument in favor of rigidity from above on employment expectations when modeling the choice of households in the labor market.

Thus, the solution to the employee household problem is the optimal level of workload e^e, which for the consumer household problem becomes an exogenous constant, while in the firm's problem it remains a phase variable. Let's return to the budget constraint and the problem of intertemporal choice of the consumer household. Let us take advantage of the fact that all variables of the budget constraint (13), except for Kln, are now exogenous constants for households, and simplify the notation:

$$\alpha\left(e^e, w, t_l, L\right) = e^e w(1 - t_l)/p^c \tag{17}$$

$$\beta\left(e^e, p^c, C_e, C_0\right) = C_e\left(e^e\right)^2 - C_0 \tag{18}$$

Then:

$$C_n(Kl_n, E_n) = Kl_n\alpha - \beta - E_n\frac{p^e}{p^c} \tag{19}$$

In turn, human capital changes under the influence of investment:

$$K_{n+1}^l = K_n^l(1 - A_l) + \frac{f_l}{L_n}E_n + \frac{f_l}{L_n}E_n^g \tag{20}$$

Since an individual uses simple nominal expectations about the state of human capital markets and changes in public policy, for him human capital is a function of one variable: E_n. Through simple mathematical transformations, we obtain a change in human capital $\Delta K_{n+\Delta n}^l$ чover the period Δn due to additional investments in it ΔE_n in the period n:

$$\Delta K_{n+\Delta n}^l = \frac{f_l}{L_n}*\Delta E_n(1 - A_l)^{\Delta n-1} \tag{21}$$

This allows us to reduce the problem of intertemporal choice to a choice E_n, for a given human capital of the first period K_n^l. Obviously, more consumption in the period n means less consumption in period n + 1. However, consumption in period n + 1 has less value for the household. Accordingly, the problem of intertemporal choice of a household is to allocate such a part of the budget for human capital in the current period so that it maximizes the total utility.

Consider the consequences of allocating additional ΔE_n for investment in human capital with a step of $\Delta n = 1$ (22). Simple mathematical transformations (23) allow us to determine that consumption in the current period is decreasing (24).

$$C_{n+1}(E_{n+1}, E_n) = K_{n+1}^l(E_n)\alpha - \beta - E_{n+1}\frac{p^e}{p^c} \tag{22}$$

$$\Delta C_n = -\Delta E_n \frac{p^e}{p^c} \tag{23}$$

$$\frac{\Delta C_n}{\Delta E_n} = -\frac{p^e}{p^c} \tag{24}$$

Consumption in the period $n + 1$ increases (26), as in the subsequent periods (27):

$$\Delta C_{n+1} = \Delta K^l_{n+1}(\Delta E_n)\alpha \tag{25}$$

$$\frac{\Delta C_{n+1}}{\Delta E_n} = \alpha \frac{f_l}{L_n} \tag{26}$$

$$\frac{\Delta C_{n+\Delta n}}{\Delta E_n} = \Delta K^l_{n+\Delta n}(\Delta E_n)\alpha = \alpha \frac{f_l}{L_n}(1 - A_l)^{\Delta n - 1} \tag{27}$$

Since the consumption of the following periods is less significant for the household than the current consumption, the effect of investments in human capital on the utility of the household after simple transformations is reduced to the formula (29):

$$\Delta U^c_n = \sum_{n=n_0}^{\tau} \Delta C_n z_n = -\frac{p^e}{p^c}\left(\frac{1}{1+r}\right)^0 + \alpha \frac{f_l}{L_n}(1 - A_l)^0 \left(\frac{1}{1+r}\right)^1$$
$$+ \alpha \frac{f_l}{L_n}(1 - A_l)^1 \left(\frac{1}{1+r}\right)^2 + \ldots + \alpha \frac{f_l}{L_n}(1 - A_l)^{\Delta n - 1}\left(\frac{1}{1+r}\right)^{\Delta n}, \ \Delta n = \tau \tag{28}$$

$$\frac{\Delta U_c}{\Delta E_{n_0}} = -\frac{p^e}{p^c}\alpha \frac{(1+r)^{\Delta n} - (1 - A_l)^{\Delta n}}{(1+r)^{\Delta n}(A_l + r)}\frac{f_l}{L_{n_0}} \tag{29}$$

Let's simplify the notation by grouping all the constants (31):

$$\gamma = \alpha \frac{(1+r)^{\Delta n} - (1 - A_l)^{\Delta n}}{(1+r)^{\Delta n}(A_l + r)}, \ \Delta n = \tau \tag{30}$$

$$\frac{\Delta U_c}{\Delta E_{n_0}} = \frac{p^e}{p^c} + \gamma \frac{f_l}{L_{n_0}} \tag{31}$$

Then the condition for the expediency of investing a household in human capital:

$$\frac{p^e}{p^c} < \gamma \frac{f_l}{L_{n_0}} \tag{32}$$

If condition (32) is met, the household allocates all earnings to investing in human capital, except for earnings spent on compulsory consumption ($C_e * (e^e_n)^2 + C_0$). Otherwise, all earnings are spent on consumption. Accordingly, the demand in the market for services that form human capital is:

$$Ye_n = \begin{cases} \frac{w_n(1 - t_l)K^l_n e^e_n - (C_e(e^e_n)^2 + C_0)p^c_n}{p^e_n}L_n + E^g_n L_n, \ p^e_n < \gamma \frac{f_l}{L_{n_0}}p^c_n \\ E^g_n L_n, \ p^e_n > \gamma \frac{f_l}{L_{n_0}}p^c_n \end{cases} \tag{33}$$

The demand in the market for goods is:

$$Yc_n = \begin{cases} C_e\left(e_n^e\right)^2 L_n + C_0 L_n + C_n^g + I_n + I_n^g, p_n^e < \gamma \frac{f_l}{L_{n_0}} p_n^c \\ \frac{w_n(1-t_l)K_n^l e_n^e L_n}{p_n^c} + C_n^g + I_n + I_n^g, p_n^e > \gamma \frac{f_l}{L_{n_0}} p_n^c \end{cases} \tag{34}$$

Note that the proposed model of a consumer household assumes sharp jumps between investment and consumer, in the narrow sense, scenarios of household behavior. It is based on the assumption that Russian households are highly adaptable. If the identification of the model shows that these assumptions are ineffective, then the model of intertemporal household choice can easily be converted from linear to quadratic.

6 The Impact of Human Capital on the Competitiveness of the Country and the Region

The proposed model can also be used for comparative analysis of economic growth and competitiveness of national economies and regions, with two significant additions and limitations. First of all, it is necessary to take into account that human capital is more mobile than material capital. This means that there is a potential outflow of labor resources. L_n, with higher human capital K_n^l from regions and countries with less developed material capital K_n^m to regions and countries to countries with higher human capital. Due to the diminishing marginal productivity of the factors of production in the model, with insufficient investment in material capital, it may turn out that the transfer of a part $L_n * K_n^l$ to another region or country with a higher material capital allows you to create there a much larger product of labor Y_n.

The substantive reasons for the emergence of such a migration flow are considered by many specialists in labor economics. Indeed, this can lead to a loss of investment in human capital. However, the phenomenon in question is actually more complex. The model under consideration indicates that an increase in human capital leads to an increase in the country's competitiveness even if a part of individuals with an increased human capital leaves.

The fact is that within the framework of the model, the measure of the economic strength of the state is not itself Y_n, a $Y_n - C_e\left(e_n^e\right)^2 L_n - C_0 L_n$. This $L_n\left(C_0 + C_e\left(e_n^e\right)^2\right)$ – "dead weight" of consumer goods going to the depreciation of labor and ensuring a living wage. It does not increase the welfare of citizens, the economic opportunities of the state, firms, etc. It is not conducive to economic growth. If we compare in the framework of the model two countries (regions) with equal factors of production (material capital K_n^m и $e_n^e L_n K_n^l$), but in one country the greatest contribution to this work is made by L_n, and in another K_n^l, then these countries will produce the same aggregate product Y_n, but the standard of living, growth rates and the possibilities of the state budget (military, foreign policy, etc.) in the second country will be higher, because its "dead weight" is less.

That is why China, with a large GDP, has far fewer resources than the United States to promote its foreign policy. For Russia and the regions of Russia, this means that attracting labor migrants from countries with a lower level of human capital is less profitable than it follows from the assessment of their contribution to GDP, and investments in their own human capital are beneficial even despite the brain drain. Quantitative assessments of these consequences will become possible in the full version of the model in the task of the state (the choice between investments in material and human capital).

Note that the conceptualization of competitiveness we are considering is close to the concept of "GDP per capita", but still significantly differs from it. In the equation of aggregate output, the market for human capital services is fully represented in per capita form, while the market for material goods is only partially. Therefore, the numerical values of the consequences of the growth of human capital for households, firms and for the state may differ.

7   Conclusion

The article proposes a two-product dynamic general equilibrium model with material and human capital as factors of production. Since both factors of production in the model are represented as capacities that increase due to investment (and decrease due to depreciation), the model allows considering the choice between investments in material capital and human capital in order to maximize economic growth. Moreover, the choice itself is non-trivial (collective), since the decision to invest in human capital is made by households, and the decision to invest in material capital is made by firms. Accordingly, the model makes it possible to raise the question of the optimal regime of public policy, which creates the correct ratio of incentives for firms and households, leading to an optimal ratio of labor and capital in the national economy.

An analytical solution to the household problem creates sufficient conditions for further identification of the model, which is a topic for a separate publication. Directly in this article, data on the dynamics of human capital that are atypical for general equilibrium models are presented. In addition, the usual time series for identifying models of this type should be used: dynamics of GDP, dynamics of consumer spending for various purposes, dynamics of output in industries of two groups (sectors that form human capital, and all others), dynamics of investments in material capital, dynamics employment, dynamics of foreign trade prices. The dynamics of material capital is restored from a combination of the dynamics of output and the dynamics of other factors of production.

The predictive strength of the model can be unambiguously measured only after its identification and simulation of the problem of firms in the model. The dynamic part of the model, which describes the firm's task, may need to be supplemented during the research process. Let us dwell, however, in more detail on those advantages of the model that follow directly from its structure and are already visible from the results of considering the household problem.

It is shown that under the hypothesis of nominal household expectations, there is no effect of crowding out private investments in human capital by public investments in human capital. At the same time, this means that households overestimate the marginal expected utility of investment in human capital: they do not take into account the inevitable decrease in the wage rate as the effective labor supply increases (the number of labor resources multiplied by the level of employment and human capital). This result is in good agreement with the phenomena observed in the Russian education market, which usually require introducing the sectoral structure of the labor market (while this model does not need it). On the other hand, the crowding-out effect can arise in dynamics due to the contribution of state investments in material and human capital to the price level in the corresponding market.

On the other hand, it is shown that within the framework of the model, the key role in the intertemporal choice of a household within the framework of the proposed model is played by the ratio of the price level in the sectors of the economy that create human capital and in the sectors of net consumption. This is a fairly strong statement, the verification of which in simulation modeling of the dynamics of the model will provide significant information about the predictive strength of the model.

The second important advantage of the model is the asymmetry of changes in the level of employment, the number of labor resources and human capital. The presence of a "dead weight" of consumption used to amortize a person leads to the fact that the total output actually distributed by firms and households is less than the total, and an increase in the number of labor resources and their level of employment increases the "dead weight", while an increase in human capital does not. This effect is useful in analyzing the consequences of interregional and international migration; it reflects the contribution of human capital to the competitiveness of a region or country.

A feature of the proposed model is the absence of a financial market in it, which is quite rare for general equilibrium models. This can be considered an advantage of the model, since the relatively low level of development of the Russian financial market and serious institutional differences of this market in Russia from the financial markets of developed countries are a serious limitation of the practical applicability of most general equilibrium models in Russian conditions. The financial market in general equilibrium models makes it possible to pose the problem of the intertemporal choice of a household between today's and future consumption (which increases due to an increase in income from bonds purchased by the household).

In Russian conditions, when the majority of the population does not have any savings (especially in bonds), does not know how to use these tools, as well as lending instruments (despite its rapid growth - however, many times lagging behind developed countries), conceptualizing the problem of intertemporal household choice vulnerable to criticism. Setting this task through investment in human capital (in the markets of which the majority of the population is involved) can open up new opportunities for explaining household behavior. An additional advantage is the direct link between investment in human capital with the competitiveness of the economy and economic growth, which is not the case for household investment in bonds.

Acknowledgments. The reported study was funded by RFBR, project number 19-29-07125mk.

References

1. Mincer, J.: Human Capital Responses to Technological Change in the Labor Market (1989)
2. Mincer, J.: Investment in human capital and personal income distribution. J. Polit. Econ. **66**(4), 281–302 (2020)
3. Becker, G.S., Murphy, K.M., Tamura, R.: Human capital, fertility, and economic growth. J. Polit. Econ. **98**(5), 12–37 (1990)

4. Black, S.E., Lynch, L.M.: Human-capital investments and productivity. Am. Econ. Rev. **86**(2), 263–267 (1996)
5. Bourdieu, P.: The Forms of Capital. Readings in Economic Sociology (N. W. Biggart, ed.), pp. 280–291. Blackwell, Malden (2002)
6. Coleman, J.S.: Social capital in the creation of human capital. Am. J. Sociol. Suppl. **94**, 95–120 (1998)
7. Schultz, T.: Investment in human capital: the role of education and of research. N.Y. (1971)
8. Aina, C., Pastore, F.: Delayed graduation and overeducation. A test of the human capital model versus the screening hypothesis (2012)
9. Crook, T.R.: Does human capital matter? A meta-analysis of the relationship between human capital and firm performance. J. Appl. Psychol. **96**(3), 443 (2011)
10. Currie, J., Almond, D.: Human capital development before age five. Handbook Labor Econ. **4**, 1315–1486 (2011)
11. Ployhart, R.E., Moliterno, T.P.: Emergence of the human capital resource: a multilevel model. Acad. Manag. Rev. **36**(1), 127–150 (2011)
12. Radaev, V.V.: The concept of capital, forms of capital and their conversion. Soc. Sci. Modernity **2**, 20–32 (2001)
13. Smirnov, V.T., Soshnikov, I.V., Romanchin, V.I., Skoblyakova, I.V.: Human Capital: Content and Types, Assessment and Stimulation, p. 513. Mashinostroenie-1, Moscow. OrelGTU, Orel (2005)
14. Galor, O., Tsiddon, D.: The distribution of human capital and economic growth. J. Econ. Growth **2**(1), 93–124 (1997)
15. Galor, O., Weil, D.N.: Population, technology, and growth: from Malthusian stagnation to the demographic transition and beyond. Am. Econ. Rev. **90**(4), 806–828 (2000)

Digitalization of Technology Transfer for High-Technology Products

Olga Pyataeva[1]([✉]) [ID], Liliya Ustinova[1], Maya Evdokimova[1] [ID],
Anna Khvorostyanaya[2] [ID], and Artyom Gavrilyuk[2] [ID]

[1] Russian State Academy of Intellectual Property, 55a, Miklukho-Maklaya str.,
Moscow 117623, Russia
o.pyataeva@rgiis.ru
[2] Moscow State University, 1, Leninskie Gory str., 119991 Moscow, Russia

Abstract. The use of digital technologies in the development of a new product makes it possible, in an accelerated time interval, to carry out information research, analysis of competitive developments and the formation of a business model, conditions are created for the intensive development of the country's economy. Key drivers of the innovation economy include knowledge, resources, human capital, knowledge databases. The constant need to innovate is dictated by the need to improve products, develop and adopt new technologies and conduct research and development. Innovative technologies play a major role in the development of production. For technological leadership, enterprises need to undertake: intensive research prior to technological development, create intellectual output and transfer technology. Information plays a bridging role in the production and promotion of goods. The paper discusses the processes by which digital technology influences the development of high-tech products. Intellectual automation with artificial intelligence (AI), in-depth analytics and big data (Deep Learning and Big Data), new tools of business modelling are considered. Advanced Analytics (advanced analyst) works with large amounts of data to help businesses make profitable strategic decisions. This approach makes it possible to identify product weaknesses and growth points, predict trends and predict potential events. The correct implementation of advanced analytics systems provides quick payback and competitive advantage in the market. In the high-technology manufacturing sector, research and development play a leading role in innovation. Software products are used at all stages of the innovation process to support the analysis and development of advanced technologies.

Keywords: Digital technologies · Knowledge bases · Technology transfer · High-tech developments · Innovation infrastructure · Development mechanisms

1 Introduction

Constant exchange of ideas and sharing of data make the innovation process effective and continuous. The relevance of the study lies in justifying the performance of innovative enterprises that actively use digital technologies. Based on the analysis of changes

A. Gibadullin (Ed.): DITEM 2021, LNNS 432, pp. 15–26, 2022.
https://doi.org/10.1007/978-3-030-97730-6_2

in the competitive environment, flexible business models are created, which are aimed at solving the problem of changes in the demand for products, on peculiarities of customers' requirements. The interconnection of the resource, organizational and economic, procedural and infrastructure blocks makes it possible to assess the situation and take high-quality management decisions.

The aim of the research is to analyse digital technologies and their role in the creation and development of high-tech products.

The object of research is innovative enterprises that actively use digital technologies. The subject of the study is the organizational and economic relations arising from the use of cross-cutting technologies in the system of strategic development of industrial enterprises.

2 Materials and Methods

The theoretical and methodological basis of the research was the work of domestic and foreign scientists on the management of innovation activities of industrial enterprises based on digital technologies.

3 Results

TT is a «transfer, transition» of intellectual property results, created in one sphere or field of activity (on one enterprise) - in other spheres and areas (on other enterprises)».

«Technology» is the sum of knowledge about the ways and means of carrying out the processes in which the qualitative change of the object takes place»; «the sum of means, processes, operations, methods, techniques, working mode», its key feature (in addition to the existence of «knowledge as an intellectual product») is the negotiability (i.e. transferability); this attribute implies, on the one hand, the existence of IP rights and, on the other hand, the material carrier in which they are expressed (on which). Thus, the objects of TT are, first, the rights to protected RIDs and RIDs (Russian Civil Code, art.), and second, the material carriers in which they are expressed (on).

Innovative technologies have another fundamental difference from traditional technologies: they are sets of methods and tools supporting the stages of innovation implementation», their ultimate goal is to bring innovation to the market.

The process of creating (innovative) technology includes: 1) creating (non-negotiable) intellectual property results (intangible) and providing it with legal protection. The rights to the created intellectual property results and material carrier in which it is expressed, but not the intellectual property results itself, may pass from one entity to another; 2) the creation of an innovative product, the embodiment of intangible objects (IP) into tangible objects and their transfer to the product market. The treatment of technology as a transferable object implies obligatory registration of rights (developer, creator, author) to it, and further: a) entry into the market with IP rights and/or b) entry into the market with material carriers in which it is expressed.

The effectiveness of the innovation development of the industry, as rightly noted by leading scientists and researchers, depends not only on the productivity of the individual innovation actors, but also on the quality of the relationships between them. The existence of a developed and extensive network of contacts between large and medium-sized companies, small firms, scientific centres, universities, authorities, non-profit organizations, etc. ensures, supports and stimulates the emergence of new ideas, Generation and dissemination of knowledge, realization of technological capabilities; expectation of increased efficiency of transfer (circulation) of knowledge, level of innovation, sensitivity of economic agents to knowledge and technologies.

Key players in TT: 1) technology developers (universities, research institutes, small and medium-sized innovation organizations, individual developers); 2) recipients (implementers); 3) intermediaries (development institutions, representatives of innovation infrastructure, venture capital market). Accordingly, the rights to intellectual property results may belong to: the Employer of technologies; the State represented by the State Customer; the Development Authority; natural persons (authors, inventors); and mixed (joint) property.

(1) The development, design of technologies, i.e. the «launching» of TT mechanisms ensure: a) the organization of the university sector (universities), which organise the execution of research and development in divisions, departments, etc. and b) Laboratories, centres of higher education consisting of scientific organizations and enterprises, scientific and educational centres (NIIs) which have been established to organize design work, to carry out scientific research for the market sector (enterprise / organization).
(2) Recipients, i.e. consumers of technology, are enterprises, organizations of the real sector of the economy, business structures.
(3) Producers and consumers of new knowledge in the real economy act as intermediaries in TT chains: a) Specialized structures (scientific centres, universities) directly involved in research and development and providing the economic agents with scientific knowledge, Scientific and technical results (in the form of patents for inventions, know-how, ready-made technical solutions, standards, etc.) and other necessary information; b) organizations collecting information on the prospects of introducing innovative processes into economic practice, which are direct or indirect clients of research and development and generate real demand for new knowledge.

Interoperability in the HHI/NII chain (1) - intermediaries (2) - enterprise (3)» appears to be a key aspect of a successful TT, and its small effectiveness is obviously a consequence of the following reasons:

Lack of conditions/capacity to develop technologies (innovative products) in the sector (1).

Lack of conditions/capacity to adopt technologies (market-oriented innovation) in the sector (3).

Lack of coordination between (1) and (3), inefficiency of the sector (2).

Next, for each industry, an analysis will be made of the key players in the TT chain and the relationship between them.

4 Discussion

The development and production of new products becomes a priority of the strategy of any firm, as it determines all other directions of its development. The use of innovative technologies makes it possible to accelerate production growth and increase business competitiveness. Equipping the production facilities with the latest equipment makes it possible to improve the quality and technological characteristics of the manufactured goods. High-tech companies are those for which information technology and information space are the main means of distribution, exchange and commercialization of the products created. In the high-technology manufacturing sector, research, development and technology, experimental production plays a leading role in innovation. Cross-cutting technologies are seen as key scientific and technical areas to ensure the competitiveness of the Russian Federation in the digital services market and industrial production [1, 2].

The list of Cross-Cutting Technologies includes promising technologies that can provide a technological advantage in the world market in the long term. Through Technology «New Manufacturing Technologies» is a complex of multidisciplinary knowledge, advanced knowledge-intensive technologies and systems of intellectual know-how formed on the basis of the results of fundamental and applied research, transfer of technologies.

Modern technologies allow machines not only to perform automatic actions, but also to interact with each other in different spheres of enterprise work. The introduction of digital technologies into the cross-cutting process involves not only production but also organizational and financial activities. The new approach ensures mobility, increasing the speed of decision-making and increasing the variability of processes depending on the needs of the client. Big data technology is one of the most important technologies for modern production. The number of data generated by production increases many times each year.

Large industrial and medium-sized businesses faced challenges in obtaining the necessary information to meet their work needs in 2020 [3].

The digital economy is a priority for most countries. A digital enterprise is an enterprise that is adaptable to present conditions and has prospects for future development. Key changes in business models through the use of digital platforms have increased the competitiveness of companies. The digital platform is a key tool for the digital transformation of traditional industries and markets, it is a high-technology business model that creates value by facilitating exchanges between two or more interdependent groups of participants. It ensures mutually beneficial interactions between third-party producers and consumers. The digital platform represents a system of algorithmed mutually beneficial relationships of a significant number of independent participants in the industry, implemented in a single information environment, resulting in a reduction of transaction costs [4]. The most important competitive advantage of digital traders in new developments and products becomes the ability to analyse and model the market: demand, supply, costs on the basis of the platform.

The digital platform is a high-technology communication platform on which the whole complex of economic relations between economic entities is realized.

Digital platforms are used in the development of new devices and technologies:

1. Industrial Enterprise Development Platform.
2. Platform for the promotion of products on the external market and the increase of export volumes.
3. Platform for analysis and forecasting of production development on the basis of objective statistics.

The technology platform is a communication tool for interaction of state, business, science for formation of program of strategic research and innovative development. The technology platform improves the conditions for the diffusion of advanced technologies in the economy and allows for the expansion of technological upgrading and the improvement of its efficiency. The effectiveness of the technology platform tool is determined by the focus on solving specific business development problems and the openness to new entrants by the flexibility of business processes (see Fig. 1).

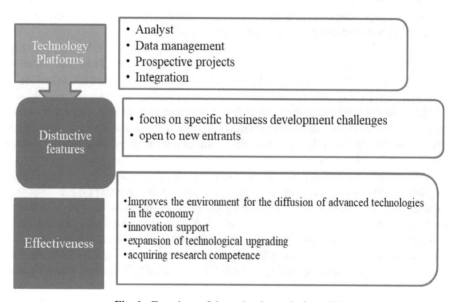

Fig. 1. Functions of the technology platform [5].

Digital transformation requires an in-depth analysis of production processes and a systematic approach to nodes, parts and assembly elements. The transition to digital production requires staff to adapt to the new work environment.

Systematizing the information obtained will help to transform knowledge into new results [5].

There is a need for an appropriate level of production and technology base, research and development, personnel skills and other important factors in the development of high-technology production. The introduction of new technologies is based on the objectives of a comprehensive increase in the efficiency of production and the creation of products required by the market. The success of innovative enterprises depends on the availability of resources and a sound technological base. Information Economy Strategy (Information Economy Strategy).

The platforms, systems and supply chains that underpin the digital market require a longer-term investment and innovation strategy and broad cross-sectoral cooperation.

Widely used types of digital production:

Clover Smart Maintenance (Data Mining Platform) - assesses the current state of plant equipment and working transport. It helps to control production quality, detect problems in time and send equipment for repair or maintenance. A modern way of processing and systematizing information, which will help transform experience into new possibilities.

HR Software (Human Resources Management) is a human resources management software. Automates the selection and evaluation of employees, calculates the system of rewards and motivations, identifies current problems in the labour market. It allows to find efficient and economical models of decision-making, to organize effective control of personnel, to anticipate problems in the future.

CRM (Customer Relationship Management) is a subsystem in the SAP system that automates the interaction between the authority and the customer. The main objective is to increase the level of sales by collecting information and analyzing interactions in order to improve the quality of service.

SCM (Supply Chain Management) supply chain management with the goal of reducing production costs and increasing production profits. SCM is a logistics technology that allows it to optimize the allocation of resources related to material or information management [6–8].

Artificial intelligence will help to manage production, control quality, reduce design time and waste, improve product reuse, and perform preventive maintenance. This will manifest itself in a rapid increase in labour productivity, optimization of possible costs, rapid adaptation to changes in the market, efficient promotion of goods and services.

ECM (Enterprise Content Management) - Corporate Content Management is a suite of applications designed to create a single enterprise information space. The main purpose of the ECM system is to maintain the full life cycle of information, from creation to obsolescence [9].

The PLM system is the place where all information about a product is combined and from where it comes in a form suitable for production and further support.

They distinguish the most advanced manufacturing sectors according to the degree of technological efficiency: production of radio and telecommunication equipment, precision and optical instruments, aerospace equipment, medical equipment. In the radio and telecommunications industry, the situation changed dramatically during the period under review. The two early-century leaders, Japan and the United States, seriously lost ground to China. China's average annual growth rate was 122% [11].

The modern economy must be based on the rapid diffusion of new technologies. Flexible policies need to be adopted in response to changing market situations, to develop different approaches to deal with particular issues. These technologies are cross-cutting in terms of universal industry use as well as efficient management and processing of the rapidly growing volume of unstructured data.

Global automation and standardization of all business processes - production, education, health and social - in recent decades has significantly accelerated the transition to digital technology.

Intelligent automation using artificial intelligence uses principles of modularity, continuous improvement, adaptive innovation. Intelligent automation can be achieved by integrating machine learning and artificial intelligence into robotic process automation to automate repetitive tasks with an additional layer of human perception and prediction. Robotic process automation is a system that can perform a certain set of tasks repeatedly and without failure, because it can be programmed to perform this type of work.

Integration of machine learning and artificial intelligence with traditional programs is required to achieve intellectual automation in order to remain competitive and efficient. A system for automating business processes in data transmission forms a centralized information hub, used at all stages of project execution for digital design, in a digital product supply network. Information hubs are a special class of online information systems whose main function is to collect and integrate multiple flows of information from different sources; to analyse, process, select and structure information; and then distribute and provide processed and best quality information to different categories of users in a convenient way.

For example, the International Technology Hub - a center of innovation - will appear in Petersburg. In Saint Petersburg, in Smolny signed a trilateral agreement between the city government, «Agency for Technological Development» and the company «Gazprom Oil». Today «Gazprom oil» is the base taxpayer and employer of the northern capital. Next year, the company plans to open a digital house, a major technological space for artificial intelligence and robotics developers. The hub will work in the spheres of management of Gazprom and solve issues of oil and gas sector. The document provides for the involvement of engineering units of industrial and energy corporations in the city, as well as support of local startups and design teams. Scanning and 3D modeling technologies are prime examples of successful digitization of products and technologies. They contributed significantly to the development of engineering, automobile and medicine.

5 Conclusions

The application of digital technologies to high-technology products has made it possible to successfully achieve such objectives:

- Real-time data management with full access;
- To create powerful and flexible systems and management tools for the creation of an innovative product.
- The widespread application of models is based on a continuous process of innovation. Monitoring and continuous optimization of processes according to the latest trends of digital transformation.

The use of artificial intelligence (AI) in innovative enterprises is high, but requires training of employees and qualitative changes in the planning and design of production. For example, the introduction of an emergent AI that adapts to external conditions, which is dynamically adaptable to specific tasks, represents a technology that makes it possible to automatically create new types of content or to modify existing ones.

Knowledge-intensive Multifunctional technologies have a wide range of applications. The main technological trends in the digital transformation of the economy and its industries include the massive introduction of the concept of the Internet of things, the active use of artificial intelligence and big data technologies, Robotization, automation and integration of production and management processes into a single information system (digital intellectual platform).

The accelerated introduction of digital technologies in the economy and the social sphere will create conditions for high-tech business, increase the country's competitiveness in the global market, strengthen national security and improve the quality of life of people. The digital platform combines all internal and external, online and offline processes of the company into a common virtual space. This makes it possible not to lose sight of the information on the state of the whole technological cycle, of service services. The management process is becoming much simpler and more visible.

The following positions should be noted as general directions of State policy.

1. Introduce a risk-based approach to the development of methodologies for allocating public funding for research, development and innovation, in particular:

– Creation of a system of risk assessment of innovative projects;
– Instituting administrative and (or) criminal investigations into possible financial abuse in the innovative sector only when an independent expert has concluded that the risk management of projects is inadequate or is manifestly inadequate;
– Creating and launching mandatory training programmes for public servants on risk management and risk-based approach;
– Formulation of methodological recommendations and norms that help to reduce excessive control and reporting in research, development and innovation organizations on the basis of a comparison of the probabilities of various abuses.

2. Formation of «trust institutions» between the innovation community and inspection bodies, in particular:

– Joint system work of innovative organizations, Regulators and legislators to improve regulatory and legislative frameworks and enforcement practices, with the primary objective of accelerating and simplifying the procedures for obtaining financing and reporting for innovative projects, based on a risk-based approach;
– Normative and legislative enshrining of the «right of an employee to error» within the framework of the decision-making powers assigned to him, in order to avoid long-term harmonization of innovative projects within the management hierarchy of organizations engaged in innovation activities, which is detrimental to the dynamics of innovative development in the Russian Federation as a whole.

Development of a limited number of simple, unambiguous, easy-to-evaluate and universally binding evaluation criteria for different types of innovation projects submitted to the public sector.

3. Prioritization of economic efficiency while developing mechanisms of social support for the released labour force, in particular:

- To introduce mandatory co-financing from the budgets of the Ministry of Labour in the regulations on modernization of production:
- Compensatory programmes for the professional staff who are released as a result of modernization;
- Retraining Programme for Personnel Released (Professional Development and New Occupations);
- Early Retirement Program (cf. for example Invalid Source specified);
- Program of financial support and for example Invalid source specified.), licensing the intellectual property of modernized enterprises to the releasing personnel for the development of innovative companies (spin-off mechanism).

Through the Ministry of Economic and Social Development, in order to eliminate the negative effects of efficiency programmes in EPA and mono-cities, establish a system of technological business incubators linked to private venture funds and mandatory State co-financing, with the possibility of licensing State-owned intellectual property to citizens.

4. Transition from the initiation and support of predominantly infrastructure projects to projects that concentrate highly professional human capital, in particular:

Introduce into the system of criteria for the evaluation of infrastructure projects in the field of innovation (technology parks, engineering centres, etc.) criteria related to the comfort and attractiveness of such projects to highly qualified professionals and their families, in particular:

- Accessibility of the chosen location of the innovation infrastructure facility;
- Availability of a sufficient quantity of quality housing in the vicinity of the facility, or plans for its construction;
- Schools, kindergartens, recreational organizations, sports complexes, catering organizations (cafes, restaurants, etc.);
- Assessment of the environmental and recreational component of the project (green areas, parks, forest stands, etc.);
- Quality medium design solutions;
- Develop and launch programs to attract innovative projects and teams from foreign countries.
- Develop and implement programmes to facilitate the acquisition of visas, residence permits, work permits and study permits for specialists of priority professions and competences for innovation, as well as for students wishing to continue their studies in these priority professions in the Russian Federation.

5. Reallocation of incentives (including financial) to stimulate demand for innovation rather than supply, in particular:

- Introduce, by reducing other taxes, mandatory contributions to central sectoral funds for modernization and innovative industrial development, supporting the search, testing, testing and introduction of innovative technologies on the orders of enterprises;

– To organize the topics of applied R&D competitions under Federal Target Programs in accordance with the orders of enterprises and co-financing from sectoral funds of modernization and innovative development.

6. Make it mandatory for organizations of any level issuing a regulation to accompany the document with explanations understandable to the target audience to which it is primarily directed, along the following lines:

– Which regulates;
– What can be done;
– What cannot be;
– What duties;
– Etc.

7. Towards better communication and building trust between industry and other actors in the innovation ecosystem:

– Oblige innovation units of industrial enterprises with State participation and add mandatory proactive research (either in-house or with outside expertise) to the job descriptions of the employees concerned Technological solutions developed in universities, science centres, innovation infrastructure facilities, as well as competence of development teams in these organizations, and CPIs in terms of number of searches and number of contacts with developers;
– Create and launch a program of specialized All-Russian sectoral conferences on exchange of information in the field of technological needs of companies and existing capabilities of domestic developers.

8. With regard to improving business models for governance and innovation financing:

– Develop recommendations on the use of modern methods, including automated ones, for business process planning of innovative enterprise structures with public participation (Strategic map method, balanced KPIs, risk-based management, etc.), with mandatory For the industry, outsourcing of change management and business process design on a competitive basis;
– Introduce recommendations on maximum «outsourcing» of innovation activities for enterprises with state participation with the contract of the relevant companies and specialists.
– With regard to changing existing practices and «traditions»:
– Further development of «cooperative mechanism» Federal Research and Development Target Programs with obligatory involvement of industrial partners and professional technology brokers.
– Introduce transparent tax incentives and State risk insurance for industrial enterprises introducing the development of small and medium-sized technology companies, universities and scientific centres.

With regard to increasing motivation for innovation:

- With regard to increasing motivation for innovation:
- Develop and recommend organizational and financial mechanisms to motivate staff to innovate (organization and development of spin-off companies, formation of joint developer companies by employees of enterprises and third party developers, financial bonuses in the form of a certain percentage of the economic impact of the introduction, etc.);
- With regard to improving the regulatory framework, accountability, legislation and enforcement;
- Develop and recommend transparent and risk-tolerant investment financing mechanisms for R&D, in addition to existing «costed" mechanisms and for partial replacements;
- Prepare and recommend for implementation a package of measures to improve the accounting policies of enterprises using an ideology of opportunity costs.

With regard to competence:

- Prepare and implement recommendations for innovation infrastructure facilities to develop activities in technology brokerage and technology scouting, as well as reverse engineering for the benefit of industrial companies. In particular, encourage innovation infrastructure facilities to form partnerships in this area with private techno-broker companies, small and medium-sized engineering companies and integrators, individual technology brokers and consultants.
- Develop and implement recommendations for calculating the cost of innovation, with mandatory costs (for example, up to 10% of the cost of external developers) paid to process brokers.

References

1. Abdrahmanova, G., Vishnevsky, O., Gochberg, L.: What is the digital economy? Trends, competencies, measurement. In: XX April International, pp. 9–12. HSE, Moscow (2019)
2. Alexeev, D.: Innovation policy of the Russian Federation: prospects for the development of the industrial sector, taking into account the application of cross-cutting technologies. Law Policy 6, 167–170 (2019)
3. Babkin, A.: Strategic management of the development of the digital economy based on smart technologies. Polytechnic-Press, Saint-Petersburg (2021)
4. Bundesministerium für Wirtschaft und Energie. https://www.de.digital/DIGITAL/Navigation/DE/Home/home.html. Accessed 21 Oct 2021
5. Chromova, A.: The digital transformation of the European industry will affect everyone.http://gosrf.ru. Accessed 21 Oct 2021
6. Chzhan, S.: The digital economy: a new growth transformation of China. http://ru.theory china.org/xsqy. Accessed 21 Oct 2021
7. Dyatlov, S.: Digital transformation of the economies of the EEC countries: development priorities and institutions. Challenges Mod. Econ. 3, 18–21 (2018)

8. EEA Digital (Internet) Trade Development Report. Eurasian Economic Commission, Moscow (2019)
9. Geliskhanov, I., Yudina, T., Babkin, A.: Digital platforms in the economy: essence, models, trends of development. Sci. Techn. State. Russian. Feder. Econ. **11**, 6 (2018)
10. Kostin, K., Berezovskaya, A.: Modern digital economy technologies as a driver of growth of the world market of goods and services. Econ. Relat. **2**, 455–480 (2019)
11. Mezentseva, O.: Development of high-tech production in the world and in Russia. Fundam. Res. **7–1**, 176–181 (2015)
12. TAdviser, Naumen: Information management practice in Russian companies. State. Bus. IT **03**, 24–27 (2021)
13. Ustinova, L.: Sustainable development of industrial enterprises and complexes in the face of external challenges. In: Sustainable Development of Digital Economy and Cluster Structures: Theory and Practice, pp. 15–20. SPBGU, Saint-Petersburg (2020)
14. Xinhong, Z.: Digital Economy: China's New Growth Variable. http:/ru.theorychina.org/xsqy_2477/201701/t20170111_349538.html. Accessed 21 Oct 2021

Reducing the International Risks of Russian Industrial Companies Based on the Transfer to the IBM Planning Analytics Platform

Ekaterina Zinovyeva(✉), Margarita Kuznetsova, Natalia Kostina, and Anastasiya Vasilyeva

Nosov Magnitogorsk State Technical University, Magnitogorsk, Russia
ekaterina_7707@mail.ru

Abstract. In modern conditions, all enterprises, including industrial ones, set the task of achieving a stable financial position in order to ensure long-term survival. The activity of a firm in all its forms is associated with numerous risks, the degree of influence of which on the results of this activity of the firm is quite high. The risk arises in any type of entrepreneurial activity related to the production of products, goods and services, their sale, commodity-money and financial transactions, commerce, as well as the implementation of scientific and technical projects. The IT industry has proven to be the most exposed to geopolitical risks. But, despite the negative consequences of the introduction of bans, domestic companies are adapting their activities to the new economic conditions. In PJSC "Magnitogorsk Iron and Steel Works" since 2019, the formation, approval and adjustment of the budget was carried out on the basis of the Anaplan automated budgeting system using cloud computing technology (SaaS), that is, the budget model is stored on servers located abroad, which implies some risks, such as the risks of disabling access to the budget model, as well as the risk of leakage of confidential information to third parties due to data storage on the Internet. Having compared the software products existing in the IT market, according to several criteria of requirements for the new budgetary system, as well as the cost of licenses, it is advisable to use the IBM Planning Analytics platform as an innovative software product for implementation in the budget process of PJSC MMK. Using the Delphi method, an expert assessment of the feasibility of implementing the developed proposals was carried out. The introduction of the IBM Planning Analytics software into the budget process of PJSC MMK will reduce the risk of information loss due to geopolitical risks by 55%. But the risk remains the same, since IBM, like Anaplan, is American software. But if a Russian analogue to American software appears on the IT market, PJSC MMK will consider the possibility of introducing this product into its activities.

Keywords: Automation · Budget process · Geopolitical risks · Industry 4.0 · Information risks · Competitiveness · Digitalization

A. Gibadullin (Ed.): DITEM 2021, LNNS 432, pp. 27–38, 2022.
https://doi.org/10.1007/978-3-030-97730-6_3

1   Introduction

Market relations that have developed in Russia have significantly increased the insta-bility and uncertainty of the state of the external environment, which greatly affects the successful functioning of organizations, therefore, there is an objective need to search and apply in practice non-traditional, but scientifically grounded innovative approaches to management that allow enterprises to achieve maximum efficiency of their activities in new conditions.

The relevance of the research topic is due to the increased turbulence and vari-ability of the external business environment and the need to ensure the sustainability and competitiveness of domestic industrial companies on the basis of effective risk management.

The impact of global sanctions on many businesses has been significant. It is quite difficult to find analogs for imported components, automation systems, software, digital technologies, especially when it comes to modernizing an enterprise.

Therefore, the purpose of the study is to reduce the geopolitical risks of Russian industrial companies through the introduction of new software products, namely, the transfer of the budgeting system of the industrial company PJSC Magnitogorsk Iron and Steel Works from the Anaplan model to the IBM Planning Analytics platform.

The implementation of Anaplan makes it possible to solve such important tasks for PJSC MMK as:

- Creation of a single model that links all stages of budgeting into a single logical cycle;
- Integration of budget planning with all modules of the Corporate Information System, as well as with MS Office tools for information exchange;
- Automatic generation of a package of management reporting;
- Functioning of the model based on common reference books and formats for the entire company.

Digital technologies are increasingly entering all areas of our life every day. The role of digitalization will grow and will largely determine competitiveness in the long term. An industrial enterprise that makes the most of the tools of the new industrial revolution in conjunction with risk management procedures can confidently call itself competitive and financially sustainable.

2   Materials and Methods

The theoretical and methodological basis of the study is the fundamental works of leading foreign and Russian scientists and practitioners devoted to modern methods of identifying, assessing and managing risks at industrial enterprises of the metallurgical industry [1–3].

The following methods were used in the work: the method of generalization and systematization of theoretical and practical material; expert method of risk assessment - the Delphi method, as well as methods such as: comparative, analytical.

The object of the research is the geopolitical risks of Russian industrial companies.

The subject of the research is the budgeting systems of Russian industrial companies.

The introduction of a budgeting system will allow enterprises to solve some problems, such as:

- Increasing the return on the use of enterprise resources - material, labor, financial, and others;
- Ensuring coordination of activities between the plant management and structural divisions;
- Provision and implementation of a forecasting, analysis, monitoring and evaluation system for various scenarios of enterprise behavior;
- Ensuring financial stability and improving the financial condition of the enterprise as a whole, increasing competitiveness.

An important condition for the functioning of an effective budgeting system is its use in conjunction with risk management procedures at all levels of the company's organizational structure. Integration of risk management into the budgeting system will improve the quality of management of the company's resources and processes.

The budgeting process involves the processing of a huge amount of information, therefore, for correct budgeting, optimization of this process and control of implementation, various systems for automating budget accounting are used.

The practical significance of the work lies in the possibility of using the proposed recommendations to optimize the business processes of Russian industrial companies in an unstable economic environment.

3   Discussion and Results

3.1   Enterprise Budgeting Process Automation

The budget system of PJSC MMK includes 60 functional budgets (budgets of responsibility centers), more than 700 cost centers (cost centers), more than 600 cost items and more than 1200 cost elements. Taking into account the importance of the budgeting process for achieving the strategic goals of the company, for the strategic management of the company, it is necessary to automate and accelerate budgeting operations using modern IT solutions.

PJSC MMK has a dynamic two-tier budgeting system: an annual budget is drawn up annually, 3 annual programs are drawn up on a quarterly basis (updating the annual budget taking into account actual data, the budget for the month and the adopted adjustments of the planned months until the end of the current year), monthly - budgets for the month. The preparation of the budget is preceded by the calculation of the optimal (in terms of marginal profit) production and sales program in the IT-system of volumetric planning of the JDA company. Then, in a certain sequence, the structural divisions of PJSC MMK involved in the budget process, based on reference books and standards, fill out their budget forms and pass them on to each other, and at the final stage of the process - to the management of the economy, where consolidation takes place, a set of all separate functional and operating budgets to the general budget of the company and preparation of materials for consideration by budget committees of different levels.

In 2018, PJSC MMK put into commercial operation an automated budgeting system based on the Anaplan platform.

The main purpose of the CRS implementation is to improve the quality and accuracy of planning based on a system of reference books that is uniform for the entire company. CRS tools on the Anaplan platform allow you to create different versions of the annual and monthly budgets and carry out comparative analysis for different plan options.

Among the main requirements for PJSC MMK when choosing a platform for budgeting is the ability to create algorithms for calculating and generating reports with minimal involvement of IT specialists and providing multi-user access to data with differentiated access rights and installation of patch protection.

The prerequisites for the transition from Excel to Anaplan were the following tasks:

- Firstly, the requirements for detailed planning have increased significantly,
- Secondly, the company's IT ecosystem became more complex when two major implementations were implemented at PJSC MMK - a volume planning system based on a JDA product and a system for consolidating the budgets of the Group's companies based on the Cognos TM1 product from IBM.

Anaplan made it possible to combine in one model information coming from various departments of the company involved in the formation and adjustment of the finished budget, reporting and analytical forms.

In order to form, approve and adjust the budget, PJSC MMK traditionally used MS Excel tools, which was accompanied by the standard difficulties for users of this tool. The model consolidates information from many different sources, in connection with which the data in the final document must be constantly updated, and in the event of a change in the structure of sources, the links must be corrected. In multi-user mode, conflicts will inevitably arise when several employees edit the same data, and access to a single model cannot be delimited by authority.

Anaplan allows you to create live and photo versions. Live versions contain all calculations, while for each live version its own settings are possible that determine the moment of intersection of the plan with the fact. Photo versions are created at a certain moment at the data warehouse level and do not contain individual settings - this is a snapshot of the system at a certain point in time, for example, the approval of the monthly budget.

In addition, a significant advantage of this CRS is the differentiation of access rights for participants in the budget process. Each account has:

- Individual settings for viewing or recording access to various directories/modules/versions/actions;
- Group settings depending on the specified role.

Each role has an individual workplace, on which, in addition to the reference data, entry forms and reports, any reference information can be placed - links to regulations, instructions, etc.

The tasks important for PJSC MMK, which are currently being effectively solved in the CRS on the Anaplan platform, are:

- Creation of a single integrated automated model that links all stages of budgeting into a single logical cycle: the formation of the annual and monthly budget according to different versions, adjustments to the model based on management decisions of various levels, analysis of the execution of the annual budget and the monthly budget according to different versions, including factor analysis.
- Integration of budget planning with all modules of the Corporate Information System, as well as with MS Office tools for information exchange;
- Automatic generation of a package of management reporting;
- Functioning of the model based on common reference books and formats for the entire company.

Problems that were solved when introducing an automated budgeting system:

- The problem of loss of information quality when transferring it from one source to another;
- Standardization of the cost accounting methodology at the planning stage and displaying the fact;
- Availability of uniform directories and forms in which users from various departments work;
- Automation of reporting and analytical information generation;
- Ease of development.

All this allows reducing the time of the budgeting process and improving the quality of the information provided.

Safety in all areas of work is a priority at PJSC MMK. Anaplan has reduced the threat to information security. Among the obvious advantages of the system, it is necessary to note the storage of information about all adjustments made to the system, which makes it possible to determine the impact of each operation on the entire system, as well as to quickly track and correct an error if it occurs.

Due to the simplicity of building a model that does not require specific knowledge in the field of programming, the platform allows concentrating the entire function of automating the budgeting system (from developing input forms and setting up links to presenting results in reporting forms) in the center of competence - the Directorate for Economics, with minimal involvement of IT specialists. And having a test platform allows you to experiment freely without jeopardizing the working model.

In addition, it is worth noting that using the Anaplan platform allows you not to worry about the problems of corporate information systems associated with the release of new versions, which leads to the need for a large-scale project to switch to a new version or functionality. The questions of technical support of the Anaplan platform are completely removed from the customer. All improvements and new versions of Anaplan appear simultaneously all over the world for all users. At the same time, without influencing the logic of the implemented projects. The user simply receives a notification about the new functionality and starts using it.

3.2   Implementation of the Digitalization Strategy in Modern Companies

Currently, most of the countries of the world economic community have entered the era of digital transformation. To increase the competitiveness of modern Russian companies, much attention must be paid to a new trend - the digital economy.

The industrial revolution (another name for this term is digitalization, or Industry 4.0) is a technological revolution, the emergence and beginning of the widespread use of new technical means [4, 5]. The industrial revolution completely changes the socio-economic, political relationships of mankind. This is how the first three industrial revolutions happened.

Since the end of the 17th century, with the advent of steam engines, society entered a technological era; during the 18th–19th centuries, significant changes took place: the transition from a subsistence economy to an industrial economy. The changes of this period are considered to be the first industrial revolution.

The second industrial revolution was marked by the emergence of electricity and conveyor production in the late 19th and early 20th centuries.

The third industrial revolution is associated with the emergence of computer technology and production automation in the middle of the 20th century. In all previous cases, the general principle remains - the invention and implementation of new technologies and the rapid obsolescence of old technologies and methods.

Today the world has crossed the threshold of the fourth industrial revolution or Industry 4.0, unlike the previous ones. It began imperceptibly, it is impossible to trace the point of its beginning, one can only abstractly assume that it started in 2009, when the number of objects connected to the Internet exceeded the number of people. It is predicted that in this system all means of production will be able to communicate with each other in real time, self-regulate and learn new mechanisms of behavior. For example, smart factories will be able to build production with fewer errors and work more efficiently.

According to the World Economic Forum, digitalization has huge potential for business and by 2025 can bring more than $30 trillion in revenue to the global economy. That is why Russian metallurgical enterprises, and among the leaders of Russian industrial enterprises, PJSC MMK, are actively looking for ways to use new technologies to increase the company's efficiency.

PJSC MMK today confidently uses information technology as a competitive advantage in all areas of its activity: in production, business processes, marketing, and interaction with customers.

PJSC MMK has come a long way, and today the company has developed a "Digitalization Strategy - 2025", which will allow the company to increase its efficiency by implementing an integrated approach to digitalization and introducing advanced technological solutions, including in the field of big data, the Internet of things, artificial intelligence and other technologies.

The strategy includes 98 IT projects, most of which are aimed at improving the efficiency of production processes. The rest will be related to research and improving production safety. According to Andrey Anatolyevich Eremin, Director for Economics of PJSC MMK, "a person is no longer able to cope with a huge amount of data. And data today is an asset to industrial companies. It is important to create a data management system in order to understand which ones are needed and how to use them". The average payback period for a digital project is 1–3 years, and its average cost is 36 million rubles.

The company will spend 5 billion rubles on digital projects. from 2020 to 2024, or approximately 2% of capital expenditures for this period ($4.3 billion). Until 2025, MMK expects to receive an additional $686 million (42.3 billion rubles) EBITDA from new projects. 14.7% (6.2 billion rubles) will be provided by digital initiatives, and in the future, digital projects will allow the company to receive an additional 4.5% of EBITDA annually.

Up to 80% of the effect will be provided by IT projects in production, the remaining 20% - by increasing the internal efficiency of the company and customer services.

To date, the company has completed several pinpoint digital projects. Let's take a look at some of them:

1. "RPA Robot" is a program that simulates human actions when working with computer systems, browsers and other sources of information. The software robot is transferred to monotonous routine work that does not require creative solutions and expert assessment, which can only be given by a human employee. The implementation of RPA allows you to speed up the execution of routine operations, reduce the cost of executing the process, and increase the productivity of the company. The basic principle of RPA is to extract, transform and load data through the user interface. An example of a frequent process is contract termination. The robot opens the mail and checks the received letter, remembers the subject and number of the contract, finds the contract in the electronic document management system by number, checks its status and the responsible employee, generates a report in the accounting system, mutual settlements under the contract, returns to the mail and prepares a letter to the responsible employee with the results verification and withdrawal. It took the robot one and a half minutes to complete routine work, which previously took an employee 20 min. RPA is effective for operations that are repeated regularly and often enough, are important for the company's business, require processing a significant amount of data, and require working with at least one electronic system.
2. "The system of operational scheduling of production" - a sufficiently developed mathematical apparatus, an optimizer that combines a large number of all systems, starting with the volumetric plan, which calculates the budget of the plant, and ending with mathematical models for calculating production schedules directly on the units. In fact, we have effective management of the entire production chain. The system adapts to constantly changing market conditions, to individual charts. We use the global On Time in Full indicator - just in time, in full.
3. "MMK Client" - the buyer's mobile application. Magnitogorsk Iron and Steel Works is a customer-oriented company, therefore digitalization is being channeled towards the convenience of customers. A program was created that allows customers online to see the progress in the redistribution of their order. They clearly understand how to create their stocks, how to plan their flows. And PJSC MMK can also regulate stocks at its own location, and this frees up cash flow in order to conduct operational activities. The most important thing is that the plant has become closer to the client; his working time has been reduced.

4. The seller's mobile assistant is used in negotiations, and when the seller begins to negotiate volumes and prices, he immediately has all the parameters he needs in his mobile phone, so that he understands what kind of discount the plant can give for this product. He can literally have all the information in his hands within one minute and continue negotiations more objectively and with better quality.

5. Implementation of mathematical models of production management. A mathematical model is a study that, through the language of mathematics, allows you to predict the behavior of a real object or solve a specific problem. A mathematical model for optimizing the purchase and consumption of coal raw materials was developed especially for the employees of the supply service by the group of mathematical modeling and system-analytical research. At the end of 2018, PJSC MMK saved more than RUB 500 million using this technology. Optimization model for the purchase and consumption of coal raw materials - a program for calculating the forecast and minimizing the cost of coal charge, taking into account the required quality of coke. This system is part of an integrated multilevel model of sintering and blast-furnace production. Today the model is used both by technologists in the process of determining the optimal coke quality and by MMK's supply service.

6. Acceptance of scrap metal using a quadcopter - The use of quadcopters allows you to obtain documentary evidence of the presence of garbage in the metal scrap when unloading from the vehicle. All this information is passed on to the supplier. In the future, PJSC MMK does not pay for this waste. At the moment, drones have helped PJSC MMK save more than 100 million rubles.

7. The mixer dispatch system allows you to see and evaluate at what time the mixer will be placed under a certain blast furnace, and where it is located at the moment. The capacity of the mixer becomes known, and blast furnace workers can estimate how much pig iron they can pour into it. Previously, the technological staff had to leave the control room, but now, by opening a tab on the computer, you can see this information.

8. Workstation "Sniper" is a program that itself determines, weighs, calculates and gives recommendations up to a kilogram on a set of ferroalloys. First of all, ferroalloys are saved, as well as money, labor costs and the quality of the smelted metal. The sniper system itself is the first step in digitalization, it is planned to digitize the full complex of each action, and the person will already act as an expert.

9. The system of automatic regulation of steam boilers - digital transformation is not a production, not a product; it is an approach to effective management for making competent accurate management decisions. The main task of the station is to collect blast furnace gas as much as possible, and to collect it efficiently, while saving natural gas as much as possible. Developed a theory of model-predictive regulation of the operation of steam boilers. What the system of automatic regulation of steam boilers gives: maximum consumption of secondary energy resource (blast furnace gas) by station boilers, minimization of consumption of natural gas as a purchased resource, we have an ecological effect, the modes are set in such a way as not to exceed the maximum permissible emission limit for the environment in terms of emissions from our boilers. If we look at such an indicator as the share of energy resources in the cost of production, then since 2016 this indicator has decreased by 1.4%.

10. Big Data (Warehouse of technological data). Rolling center No 11 is known for producing metal products for the automotive industry, as you know, automotive industry enterprises are the most demanding, and it is not an easy task to be suppliers to the automotive industry. The final quality is additive, and is an inherited combination of the qualities of rework pieces and ideally executed processes. A warehouse of technological data was created by a statistical method, the warehouse includes technological parameters from smelting (from the first processing) to the shipment of finished products. More than 10,000 parameters are recorded per day, they are archived, stored and subsequently you can work with them, tracking the technology, evaluating its suitability for quality assurance. Any deviations from technology require very quick and effective decisions, and this storage is a kind of foundation, a basis for making such decisions, for developing control actions.

11. Smart digital warehouse. In LPTs No. 11 at NGTs-3, the functionality of a smart warehouse is implemented at the input and output warehouse. It allows you to track the material flows of metal products, store them in a high-quality manner, make decisions for technological personnel according to the suitability that they will see in their information system both on the cranes and in the foreman's. Time is freed up to check the integrity of the package, the accuracy of loading, and also check the metal products by comparing with the shipping documentation, and will not allow the wrong roll or roll of less weight to be shipped to the consumer.

12. Adaptive technologies - within the framework of digitalization at RTOs in 2019, two main pieces of equipment were introduced into production: a 3D scanner and an adaptive 3D printer installation; casting. If the classical technology of using sand-clay mixtures allows to obtain castings of 13–14 class of accuracy, then this installation can obtain castings of 7–8 class of accuracy.

13. Mobile MRO. The main tasks of the MRO process are planning repairs, financing maintenance costs and financing common tasks. The summarizing cost is the analysis and reporting system, which allows you to find the reasons for the deviation of the main maintenance parameters from the specified ones and make competent adjustments. All equipment is entered into the database and divided into critical and non-critical. Routes for the service personnel have been drawn up, approaching the NFC tag, he reads it and sees the equipment that needs attention, he enters all the data using a mobile device into the system. When determining the causes of breakdowns, you can analyze this data and change the maintenance strategy somewhere.

14. The Industrial Internet of Things allows solving many problems associated with the inaccessibility of components and assemblies for diagnostics, as well as reducing the risks of damage to equipment, cable routes, and reducing the cost of laying them. At the 370 MMK mill, a system of vibration diagnostics sensors was used, the sensors are directly installed in the gearbox, and all data is transmitted via a closed radio channel to the control post. Among the advantages of the vibration diagnostics system are short measurement times, no stopping and disassembling of the mill equipment. Such directions are developing not only in relation to the condition of the equipment, but also in relation to the conduct of technological processes. This system is being implemented at BF No 10 and is associated with the analysis of the thermal state of the blast furnace.

15. Digitalization also affected the budget process of PJSC MMK. In 2018, the Anaplan automated budgeting system was put into pilot operation.
16. The digitalization strategy of PJSC MMK is being carried out at all levels, from sintering and blast-furnace production to logistics and accounting. This is a practical experience transferred to digital, which will increase labor productivity and quality of products, ensure the safety of the life and health of employees, and become closer to customers, their expectations and needs. In the near future, the digitalization strategy will be the main driving force and key driver of the mill's sustainable development.

3.3 Development of Proposals for Reducing Geopolitical Risks, Optimization and Unification of Business Processes in Russian Industrial Companies

To achieve the goal of reducing geopolitical risks [6], it is necessary to introduce into the budget system of PJSC MMK such a product that will be based on On-premises technology with the placement of software on its own server equipment owned by Informservice LLC.

According to experts in the field of IT technologies, the following systems are the leaders among budgeting software products:

– Anaplan;
– Oracle;
– SAP;
– IBM;
– Prophix.

We will use Gartner's Magic Quadrant to select the optimal software products for PJSC MMK.

Gartner is an IT analytics and consulting company. She researches markets, publishes reports and advises other companies. Gartner Magic Quadrant is one of their popular types of reports presented on the study of a particular IT market [7].

In its reports, Gartner considers not only the quality and capabilities of the software, but also the characteristics of the developer as a whole, such as sales and customer experience, market understanding, business model, innovation, marketing strategies, sales, industry development, etc. Based on estimates by key parameters vendors are divided into 4 groups: leaders, candidates for leadership, visionaries and niche players.

Let's interpret the Gartner quadrant as follows:

– Prophix belongs to the category of niche players, that is, solutions that are not very widespread in this market. As a rule, they are geared towards solving certain or very specific (narrow in relation to this entire market) tasks - niche solutions.
– Anaplan belongs to the category of visionaries. These include those solutions that are not yet strongly represented on the market, but actively talk about the prospects for their development in the future. Newbies often appear here with a ready-made and rapidly developing product.

– SAP - complete holistic solutions that are widespread enough in the market, but do not show clear trends for future development, a contender for leadership, provided that clear development trends are identified.
– Oracle and IBM are market leaders. A well-proven complete solution that covers the needs of the vast majority of customers and has an understandable development in the future.

Based on the comprehensive analysis carried out, it was revealed that the IBM Planning Analytics platform is the most optimal as an innovative software product for implementation in the budget process at PJSC MMK.

IBM Planning Analytics is a modeling system based on Cognos TM1, which is already used in two business processes at PJSC MMK - in the preparation of financial statements in accordance with RAS and IFRS, and in the preparation and analysis of the consolidated budget of the Companies of the PJSC MMK Group.

Key benefits of IBM Planning Analytics are:

– No risks associated with disabling access to the budget system;
– Cheaper license cost compared to Anaplan;
– The opportunity for economists to independently create and adjust any elements and relationships of the budget system;
– Ability to expand the boundaries of the project as needed;
– Modern tools for analysis and presentation of information;
– High performance characteristics and integration with other KIS modules;
– Customizable dashboards with the ability to enter data;
– Configuring workflow in the system;
– 100% functionality on any mobile device.

IBM Planning Analytics is a modern, convenient, flexible, comprehensive solution for planning, data analysis and process control within a separate structural unit or company as a whole [9].

4 Conclusion

According to the results of the study, the proposed measure for the introduction of the IBM Planning Analytics software product into the budget process of PJSC MMK was implemented in August 2020 together with the Russian contractor GMCS, and from that moment the preparation of the monthly budget was fully carried out in this system.

In September 2020, the implementation of the product at OJSC MMK METIZ and LLC OSK was completed, and the preparation of the draft budget for 2021 using IBM Planning Analytics began, in particular, the process of protecting functional budgets was automated, which was previously carried out in paper form. At the moment, the draft budget has been fully approved and signed.

For 2021, it is planned to implement IBM Planning Analytics in two more Companies of the PJSC MMK Group: LLC MRK and LLC MMK LMZ.

As part of the Industry 4.0 digitalization strategy, in July 2020, together with GMCS, design work began to prepare for the implementation of a comprehensive modeling system and optimization of equipment repair programs based on IBM business solutions. This system will make it possible to determine the most effective measures for the repair of equipment, as well as to plan the financing necessary for their implementation. This approach provides a significant increase in the efficiency of stations, and the potential economic benefit from the implementation of the system will amount to tens of millions of rubles.

Also, according to the "Digitalization Strategy 2025", within several years, the development of a project for the implementation of an automated volumetric production and economic planning system based on IBM ILOG will begin.

Thus, in the long term, a budget system will be created in PJSC MMK that will link together all the structural divisions involved in the formation, approval, adjustment and analysis of the budget of the Companies of the PJSC MMK Group.

References

1. Antonyuk, V.S., Danilova, I.V., Erlikh, G.V.: Risks of a town-forming enterprise in the risk system of a company town. In: 3rd International Conference on Industrial Engineering, vol. 35 (2017)
2. Ivashina, N.S., Kuznetsova, M.V., Zinovieva, E.G., Litovskaya, Y.V.: Developing the system of evaluating risks of metallurgy enterprises taking into account the synergistic approach. In: 3rd BEM International Conference on Education, Sociology and Humanities (BEM-ESH 2018), vol. 12, pp. 225–232 (2018)
3. Karelina, M., Ivanova, T., Trofimova, V.: Expert system for risk assessment of M&A-Projects. Int. Bus. Manag. **9**(5), 762–770 (2015)
4. Zhang, C., Chen, Y., Chen, H., Chong, D.: Industry 4.0 and its Implementation: a review. Inf. Syst. Front. 1–11 (2021). https://doi.org/10.1007/s10796-021-10153-5
5. Majstorovic, V.D., Mitrovic, R.: Industry 4.0 programs worldwide. In: Proceedings of the 4th International Conference on the Industry 4.0 Model for Advanced Manufacturing (AMP 2019), pp. 78–99 (2019)
6. Lu, Z., Gozgor, G., Huang, M., Lau, M.C.K.: The impact of geopolitical risks on financial development: evidence from emerging markets. J. Compet. **12**(1), 93–107 (2020)
7. Gartner Magic Quadrant. https://www.gartner.com/en/research/methodologies/magic-quadrants-research. Accessed on 21 Sept 2021
8. Oracle Hyperion Planning, https://www.oracle.com/applications/performance-management/products/hyperion-planning/. Accessed on 21 Sept 2021
9. IBM Planning Analytics. https://www.ibm.com/ru-ru/products/planning-analytics. Accessed on 21 Sept 2021
10. Zinovieva, E.G., Koptyakova, S.V.: Assessment of integration risks for metallurgical enterprises using the fuzzy set method. CIS Iron Steel Rev. **17**, 58–64 (2019)
11. Banno, M., Tsujimoto, Y., Kataoka, Y.: The majority of reporting guidelines are not developed with the Delphi method: a systematic review of reporting guidelines. J. Clin. Epidemiol. **124**, 50–57 (2020)

Adaptation of the Experience of Digitalization of the Chinese Insurance Industry in Favor of the Development of Technologies of the Russian Insurance Market

Wang Wentao[1] and Evgenii Makarenko[2]

[1] St. Petersburg State University, 13B, Universitetskaya Emb., St. Petersburg 199034, Russia
[2] St. Petersburg State University of Aerospace Instrumentation, 67, Bolshaya Morskaya Street, Saint Petersburg 190000, Russia
ss300@yandex.ru

Abstract. This article mainly describes the digital trend and development status of the Chinese insurance market in the 21st century. The statistical methods are used to highlight the role of digital technology in the development of the Chinese insurance market. The result of this article is that digital technology promotes the rapid development of the Chinese insurance market, accelerates the digital trend the market more and more open. The number of the companies investing in China are growing, so the development prospects of Chinese insurance market are extremely broad. The technologies of cloud computing, big data and artificial intelligence affect the development of the digital economy. The Chinese Internet companies already have got the independent innovations in the said, which has laid the foundation for the digitalization of the Chinese insurance market, which is conducive to expanding the scale of the insurance market, accelerating the pace of the Chinese insurance company system reform, and accelerating the internationalization of the Chinese insurance industry. The parallelquel of the insurance markets development of Russia and China.

Keywords: Chinese insurance market · Formation · Digitalization · Structural characteristics · Insurance company · Insurance industry · IoT · Insurance premium

1 Introduction

The 21st century is the era of rapid technological development, people's living standards have improved significantly, and the digital economy has entered the period of comprehensive development. China has become the official member of the World Trade Organization, which has steadily increased the country's openness to the outside world and promoted the overall development of Chinese market economy. The implementation of the One Belt One Road strategy has strengthened the cultural biases between China and all the other world countries, thus having improved the international status of China. Therefore, the Chinese insurance market has entered the digital age. The

A. Gibadullin (Ed.): DITEM 2021, LNNS 432, pp. 39–46, 2022.
https://doi.org/10.1007/978-3-030-97730-6_4

more economical development and social progress condition the growing demand in insurance.

2 Materials and Methods

2.1 Digital Trends in Chinese Insurance Market

More and more Chinese insurance companies are tending to use digital technology to improve the service level and accelerate the company's digital strategy development.

The digital strategy is one of the important development strategies. The company will fully implement the digital strategy from three aspects: customer experience, business operations and business models [13, 21].

The digital transformation of insurance companies means the digitization of business management as a whole. Internet technologies and artificial intelligence have effectively promoted the digitization of the insurance market [20].

The development of Internet insurance marks the digitalization of the insurance market. The Chinese insurance market is facing an unprecedented digital transformation with the rise of insurtech companies, the surge in demand for Internet insurance scenarios, and the expansion of online sales channels [14].

Currently China has been a success in IoT usage in all the insurance businesses. At that, it is noticeable that this process includes all the stages of the interaction between the insurer and the insured and the underwriting automatization using the Big Data service (Fig. 1) [5, 10].

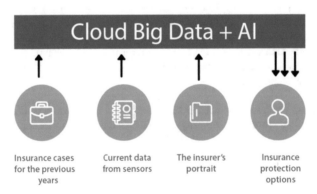

Fig. 1. Scheme of using IoT technology for personalizing insurance service.

The digitalization trend in Chinese insurance market is due to the following [9]:

- Use big data technology to improve customer demand capabilities.
- Use artificial intelligence technology to optimize marketing, claims settlement, user service and other links, reduce costs, and bring customers a better experience.
- The cloud computing platform can carry the rapidly growing mass of products and user data, and quickly realize the optimization and upgrade of the system and platform with reduced costs.

- Increase user interaction channels through IoT technology to obtain more user information.
- Improve the availability and convenience of insurance products and services through the Internet and mobile technologies, break time and space restrictions, achieve remote or 24-h customer service, obtain more information, and combine big data for product development and marketing.

The vast majority of traditional insurance companies use big data technology to develop insurance products that meet the needs of consumers, and provide insurance products and services to consumers through Internet platforms, thus realizing part or all of the insurance business networked business activities [11].

In recent years, the digital development of Chinese insurance companies has made great progress. The application of a large number of digital software and systems has broadened the insurance sales channels. It has become easier for customers to purchase insurance, and the price has become cheaper and more transparent [15]. The communication and cooperation of all parties involved in insurance have become more convenient [19].

While the insurance industry is developing, Chinese Internet economy is also developing rapidly, and Internet insurance channels have promoted the rapid development of the insurance market. Internet insurance is a symbol of the digital trend of China's insurance market. According to statistics, as of 2017, China's Internet insurance premium income fell to 187.527 billion yuan, a year-on-year decrease of 20.1% [7]. In 2018, China's Internet insurance premium income reached 188.9 billion yuan [14] (Table 1).

Table 1. Chinese Internet insurance premium income 2012–2020.

Year	Chinese Internet insurance premium income 2012–2020 (100 million yuan)
2012	97.5
2013	110.7
2014	318.4
2015	859.0
2016	223.4
2017	234.7
2018	187.5
2019	188,9
2020	202.3

From 2017 to 2019, Internet channel property and life insurance income experienced negative growth. In the first half of 2018, it fell by 20.01% and 10.9% respectively. The personal insurance Internet insurance premium income disclosed in the first half of 2019 fell again by 15.61% [6]. There are different opinions on the prospects of Internet channels, but the consensus formed in the industry is that the insurance technology evolved from the Internet technology wave will change the traditional model of the insurance industry, reshape the future of the insurance industry, and will become the next major part of the insurance industry.

In the process of digital transformation, insurance companies have the following problems [4, 21]:

– Difficult to break through traditional sales thinking;
– Lack of service awareness;
– Lack of thinking of using data to solve problems;
– The business model lacks innovation;
– The interactive platform for communicating with customers is inefficient.

To this end, the Chinese insurance industry needs a number of reform measures:

– Adjust the insurance organization according to local conditions;
– Enhance the innovation of the team;
– Build a digital insurance talent team;
– Build a digital insurance culture;
– Research and study the history of the Japanese insurance industry.

2.2 Digital Trends in Chinese Insurance Market

When considering the digitalization development on the Chinese market of financial services, one can represent it as the following scheme (Fig. 2). Personalization and eco systems development became the core trend of 2020.

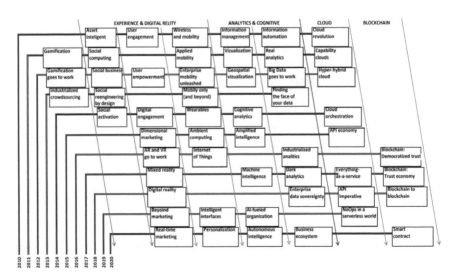

Fig. 2. Chronological scheme for the use of modern information technologies in the insurance business.

In order to promote the sustainable development of the Chinese insurance market and the digitalization trend, the Chinese government began to adjust the insurance market. China's socialist market economy is relatively developed, which makes the reform of the market system the core link in the development of the market economy.

The following aspects are to be reformed:

– Due to the rapid development of the social economy, the Chinese insurance market is facing severe challenges, it needs to be institutionalized. In various regions of China the development level of the insurance market varies greatly. The system reform is to develop market resources thus promoting the rational allocation of them;
– The Chinese insurance market must show the entrepreneurial spirit which is the fundamental driving force for the development. The obstacles to the growth of entrepreneurship in the Chinese insurance market are mainly reflected in the loss of freedom caused by improper control, and the insufficient restriction of power. The Chinese insurance market needs to learn from the development history of the American and Japanese insurance markets, and strengthen cooperation with foreign insurance markets to promote its own sustainable development [18].

It is also worth mentioning that China is experiencing the struggle with the eco systems development in order to decrease the cross sales and collecting information about users. One can assume that this trend will keep going, and the cloud technology and Big Data will follow the way of less personalization or less interpenetration of social networks or messengers into the financial systems.

3 Results

The Chinese insurance market must give full play to the role of insurance mechanisms in social governance; due to insufficient market understanding, insurance mechanisms have not received sufficient attention in social governance, which is an important factor restricting insurance market entities from deepening the division of labor and improving the level of professional management. Giving full play to the role of insurance mechanisms in social governance can not only promote the development of the insurance industry, but also promote the formation of a central social governance structure. The insurance mechanism also helps to promote the deepening of the market division of labor and the improvement of the level of professional operation, and promote the formation and development of a division of labor that conforms to market logic [1, 3].

Insurance companies should cultivate excellent corporate culture. Chinese insurance companies should strengthen cooperation with insurance companies in developed countries, learn from their corporate culture and study their modern development strategies. The excellent corporate culture of insurance companies is conducive to promoting the reform and development of the insurance market.

4 Discussion

Since the beginning of the 21st century, with the rapid development of the national economy and the advent of a new round of economic growth cycle, the Chinese insurance industry has gradually got rid of the old system and ushered in a benign, relaxed, and open economic, institutional and institutional environment. Opening up and competition are undergoing rapid changes. In the past five years, premium income has increased by more than 24% annually, 2.8 times the average GDP growth rate over the same period. At the end of 2005, the total assets of the Chinese insurance industry exceeded 1.5 trillion yuan, and the premium income that year was nearly 500 billion yuan [16, 17]. While the scale is expanding rapidly, the depth and breadth of market opening is also constantly developing, and a diversified market competition pattern is taking shape. At the same time, modern corporate governance mechanisms have been successfully introduced into the Chinese insurance companies, and the reform of the insurance system has taken historic steps.

The total assets of the Chinese insurance industry only account for about 3% of the total financial assets, while in the United States, the proportion is as high as 35% [8, 12]. This is enough to prove that the Chinese insurance market is still an immature market.

The development of the Chinese insurance market is not only required for the development of the insurance industry itself, but also for the development of the entire financial market and the development of the national economy. Facing the profound changes in the Chinese insurance market, whether it is insurance practitioners, insurance researchers, or regulators, if they want to gain the initiative in a market full of opportunities and challenges, and enhance their ability to mobilize resources in the financial market, they hope to The era of global openness of the market is increasing its strengths and avoiding weaknesses, enhancing the ability of Chinese insurance companies to deal with international competition, and promoting the stability and health of the Chinese insurance market in the process of rapid growth.

First, the market scale is expanding rapidly. The rapid growth of the market scale, on the one hand, shows that Chinese insurance market has a vast space, on the other hand, it also stimulates social capital's confidence and enthusiasm for investment in the insurance industry, and further accelerates the pace of insurance industry development.

Second, the degree of market openness is constantly increasing.

Third, the trend of diversification of market entities has further developed. With the further deepening of marketization, the main body of Chinese insurance market has increased rapidly.

Fourth, a historic step has been taken in system reform. Fifth, the business model is undergoing new changes.

The supervision and management of the insurance market by the Chinese government has played an important role in promoting the reform and stable development of the insurance market, and has enhanced the international competitiveness of Chinese insurance companies.

In Russia the information technologies market follows the Chinese way, but still it may be noted that on the one hand the demands of the regulator to the financial component are being tightened, on the other, the development of ecosystems (for example, Sberbank) is practically not regulated in any way, which potentiates the use of product customization. In this case the main obstacle is mostly the informational systems weakness, as well the poor data filling, which makes impossible to create the digital portraits of the insurer.

5 Conclusion

Since the 21st century, the reform and development of the Chinese insurance industry has made brilliant achievements and played an important role in promoting reform, safeguarding the economy, stabilizing society, and benefiting the people. Facing the future, Chinese insurance industry has huge development potential and broad development space.

The insurance market is inseparable from several key elements such as market transaction subject, transaction object and price. Market transaction subjects mainly include demanders and suppliers in the insurance market and other subjects that facilitate insurance market transactions such as insurance intermediaries and regulators. Market transaction objects are insurance products provided by suppliers to demanders.

The digital development of Chinese insurance companies has made great progress. The application of a large number of digital software and systems has broadened insurance sales channels. It has become easier for customers to purchase insurance, and the price has become cheaper and more transparent. The communication and cooperation of all parties involved in insurance have become more convenient.

Using the experience of digitalization of the Chinese market by domestic companies allows to omit the search for the development ways (which will undoubtedly lead to extra expenditures), but to use the systems developed by China, taking into consideration the indulgences that are still in domestic legislation in terms of personal data protection.

References

1. Belozerov, S.A., Chernova, G.V., Kalayda, S.A.: Modern factors of development of the Russian insurance market. Insur. Bus. **3**, 31–35 (2018)
2. Bozhuk, S.G., Maslova, T.D., Pletneva, N.A.: Improvement of the consumers' satisfaction research technology in the digital environment. In: IOP Conference Series: Materials Science and Engineering, p. 012055 (2019)
3. Chernova, G.V.: Textbook for Bachelors. Yurayt Publishing House, Moscow (2019)
4. Grebenshchikov, E.S.: The insurance market of China: rates, scales and experiments. Finance **1**, 38–44 (2017)
5. Huang, Z.: The Power of Digital Finance: Empowering the Real Economy. Renmin University of China Press, Beijing (2018)
6. Ivanova, V.V., Pokrovskaya, N.V.: The Financial System of China. Prospect, Moscow (2018)
7. Kou, Y.: Blue Book of Insurance, Analysis of China Insurance Market Development. China Economic Publishing House, Beijing (2017)
8. Liu, Z., Liu, B.: Introduction to Insurance. China Finance Press, Beijing (2017)
9. Ma, H.: Digital Economy: China's New Growth Momentum for Innovation. China CITIC Press, Beijing (2017)
10. Ma, W.: Digital Economy: Discovering New Opportunities in Traditional Industries and Emerging Formats. Democracy and Construction Press, Beijing (2017)
11. Makarenko, E.A., Pesockij, A.B.: Scoring system as a mechanism for preventing buncruptcy of insurance companies. In: European Proceedings of Social and Behavioural Sciences EpSBS, pp. 1446–1455 (2020)
12. Zhong, A.: New Insurance Era: Fintech Redefines the New Future of Insurance, pp. 25–43. Technology Finance Research Institute. China Machinery Industry Press, Beijing (2018)

13. Tang, D., Zhang, X.: Insurance Theory and Practice. Beijing Institute of Technology Press, Beijing (2017)
14. Tang, X.: Digital Economy: New Technologies, New Models, and New Industries that Affect the Future. People's Posts and Telecommunications Press, Beijing (2019)
15. Tcerkasevich, L.V., Makarenko, E.A.: Development of insurance from infectious diseases in Russia. Sci. Work Free Econ. Soc. Russ. **224**, 386–401 (2020)
16. Wang, X.: Insurance. Baodeng Education Press, Beijing (2017)
17. Xie, S.: Blue Book of Automobile and Insurance: Report on Development of Chinese Automobile and Insurance Big Data. Social Sciences Academic Press, Beijing (2020)
18. Yang, Z.: Principles of Insurance. Tsinghua University Press, Beijing (2018)
19. Zhao, Z.: Internet Insurance. Capital University of Economic and Business Press, Beijing (2017)
20. Zhu, J., Liu, Y., Wei, L.: Insurance Technology. China CITIC Press, Beijing (2018)

Processing of Streaming Weakly Structured Data

Olga Denisova$^{(\boxtimes)}$ (iD)

Ufa State Petroleum Technological University, Kosmonavtov Street, 1, Ufa 450062,
Republic of Bashkortostan, Russia
denisovaolga@bk.ru

Abstract. The article discusses systems that allow performing continuous queries, as well as their architecture. It was found that the work of such systems begins with the construction of a query execution plan. The request is executed by an executor similar to the executor used in database management systems (DBMS). The existing systems for performing continuous queries over data streams in JSON format currently do not meet the requirements. It is necessary to develop our own system for executing continuous queries over JSON streaming data based on modern technologies for streaming computing.

Keywords: Big data · Database management system · Continuous query system · Data flow · Streaming computing

1 Introduction

Modern research centers, social networks and financial institutions generate and collect data, the measurement of the volume of which is already habitually made in petabytes. For example, the data received by the large hadron collider is about 1 Pb/s, the NYSE Stock Exchange creates and replicates about 1 TB of data daily, and more than 15 billion images are stored in Facebook data centers.

This determines the relevance of the big data problem in the modern world. World leaders in the field of business and information technology are searching for the optimal solution aimed at managing and analyzing huge volumes of permanently arriving information. They are busy looking for ways to benefit from the data that is at their disposal.

It should be noted that the topic of big data is of interest not only from a practical point of view, but also from the standpoint of theory. Due to the continuous process of development of the technologies themselves, this allows, along with real-time monitoring of their implementation and improvement, also to take a direct part in the creation of new technologies used for processing large amounts of data. At the same time, it is worth emphasizing the special relevance of expanding skills and knowledge in the field of big data for students of applied informatics specialties [1–3].

In the software development industry, there is a class of tasks for which it is necessary to respond as quickly as possible to data coming from outside. It is usually necessary

© The Author(s), under exclusive license to Springer Nature Switzerland AG 2022
A. Gibadullin (Ed.): DITEM 2021, LNNS 432, pp. 47–58, 2022.
https://doi.org/10.1007/978-3-030-97730-6_5

to monitor in real time some production or business processes in which changes often occur or new data is generated, and based on these changes, perform some predefined actions. We should immediately note that in relation to such problems, the concept of "real time" usually has a weakened meaning—for systems that solve them, there are no requirements in the form of the maximum permissible reaction time to changes occurring in the observed processes. It is assumed that a long reaction time affects only the quality of the operation of such a system. In the future, we will use the concept of "real time" in this sense. The data from the observed processes is usually considered as a stream of events corresponding to the changes taking place. Such tasks are usually called streaming processing tasks, and the incoming data for them is streaming. Examples of areas where such tasks usually arise are the following:

- Stock markets and currency exchanges. During the trading process, a huge number of transactions constantly occur, each of which affects the overall situation on the trading platform. Taking into account these transactions, it is necessary to constantly update information about the general state of trading, recalculate currency rates and stock quotes, maintain various statistics, provide them to companies and traders participating in trading.
- Telecommunications networks. In the networks of Internet access providers, networks of mobile operators, tasks that require real-time streaming processing also often arise, such as monitoring the state of the network and equipment, detecting and suppressing malicious activity on the part of users, as well as collecting statistics on network usage and its load.
- Sensor networks. In such networks, data comes from a variety of sensors or devices (sensors) that periodically generate some events. For example, it can be a GPS receiver mounted on a car that sends its coordinates every minute, or an RFID tag scanner [4] at the exit of the library, which tracks all books taken out of its premises. By collecting information from a variety of sensors and processing it in automatic mode, it is possible to build systems that monitor the correct functioning and control complex processes of the real world, while practically requiring no human intervention. In modern high-load applications and web services, the tasks of streaming processing do not lose their relevance. In particular, it is necessary to solve the problems of calculating financial indicators in real time, such as the amount, the average price of transactions for the last day; identifying popular content-determining the most frequently mentioned news in recent times; monitoring the state of the service-tracking the number and types of requests that require too much time to respond to. The success of business processes and the entire companies in which they arise depends on the quality of solving such problems. Therefore, there is a whole field of systems designed to simplify the solution of problems of calculating analytics and metrics in real time.

At certain stages of the development of the technological process and society, a problem arises related to the inconsistency of the growth of information volumes and the technical capabilities of existing software products and computer equipment. In the process of human activity, information accumulates describing the objects of the surrounding world (primary description). Object models and data models (secondary description) are built on its basis. The more complex the object under consideration, the more information is needed to describe it, and this leads to the complexity and volume of information collections that make up such a description.

The search for new solutions related to software, the increase in RAM, the search for data storage capabilities, processing, filtering of information, is an urgent task, the solution of which will allow us to reach a qualitatively new level of development of science, technology and technology. The growth of data arrays is the subject of research not only by IT companies, but also by scientific and educational organizations, companies engaged in trade, business, marketing, and healthcare. Modern science identifies a separate branch that studies the analysis of large and ultra-large data arrays, in which an additional factor of research is not conventional models, but information models as a system information resource. This formulation of the question opens up new opportunities for solving a whole range of new tasks, but here a new problem is revealed: the organization of this system resource. Creating applications applied to big data is associated with some difficulties:

- Large amounts of data;
- Intensified data flows;
- The time limit for making decisions with any amount of data;
- A significant reduction in the allowable time of data analysis;
- Increasing structural complexity of systems and models;
- Increasing morphological complexity of models;
- Increasing computational complexity;
- Relative growth of fuzzy information;
- Relative growth of weakly structured source information;
- Increasing the need for parallel computing and so on.

Applications that are focused only on working with large amounts of data process information ranging from a few terabytes to 1 PB. Such data is received, as a rule, in several different formats. At the same time, there is often a distribution of this data between several locations. Traditionally, to process such data sets, a multistep analytical pipeline mode is used, which includes the stages of data transformation and integration.

In the case of an increase in the volume of data, as a rule, there is an almost linear increase in the requirements for computing. Calculations are often amenable to ordinary parallelization. Among the key research problems should be highlighted data management, methods used for data integration and filtering, as well as effective support for queries and data distribution.

At the same time, it is worth paying special attention to the distribution of data, because even in the case of a not very large volume, it creates problems in itself. This is the reason motivating the process of developing special spatial data models, often displaying the properties of a field or the properties of an information space.

2 Materials and Methods

The process of big data analysis for enterprises and organizations is fraught with a number of difficulties, which are associated with the lack of software used for information analysis, as well as personnel problems in this area. The second problem that leads to the processing of inadequately huge amounts of information is related to their quality, in addition, these volumes also need to be stored. Currently, various manufacturers offer software products that have significant technical capabilities, for example, platforms that allow them to process large amounts of data, as well as capable of solving complex technical and technological tasks. To select the processed information, it is necessary to use complex methods of data search and integration. The main research difficulties include the construction of new algorithms, filtering of generated data, as well as the creation of specialized platforms that include hardware accelerators.

Many of the streaming processing tasks can be represented as so-called continuous requests over the stream of incoming events. A large number of tasks for calculating analytics and metrics in real time can be easily expressed using continuous queries written within some special query language. Many systems provide the implementation of such languages. Next, we will consider only such systems. Let's consider their basic concepts:

An event is a unit of information corresponding to some action that occurred in the observed process, which provides interest in the task and should be processed. An event stream is a potentially infinite sequence of events that enter the system as they occur in the observed process. A continuous query is some constantly running query over events from the stream, set by the user, the result of which should be updated regularly as new events arrive. An important property of continuous requests during streaming processing is that their result should never be recalculated completely, but only partially updated when new events arrive.

Systems designed to perform continuous queries must support the means to set them. Therefore, similar to database management systems, there is a need for a high-level and expressive query language that allows you to express the required transformations on data. Unlike a DBMS, where data for queries comes from the built-in storage, when executing continuous queries, they come from the stream of incoming events.

When performing real-time analytics tasks, the following functional aspects are important, which are primitive blocks on the basis of which complex algorithms can be built to solve streaming processing problems. In a continuous query execution system, they must be expressible in the query assignment language. Let's consider these aspects:

– Access to the structure of incoming events. The events that have come can be considered simply as the fact that some action has occurred in the observed process. But usually, it can contain a fairly large amount of useful information that you need to provide access to. Therefore, the query language requires the ability to access data inside events;
– Filtering the event stream. During the execution of a continuous request, only some events from the incoming stream may be of interest. In order to filter out the rest, it is necessary to support the filtering mechanism. It is possible to filter events by type, time of receipt, based on the information contained in them;
– Building new event streams. The result of a continuous query can be either just some constantly updated value, or a stream of events, over which continuous queries can also be set. In the second case, the functionality of the system in the case of combining requests can greatly increase;

- Grouping of events and data aggregation. Calculating various aggregate values over a real-time data stream is one of the fundamental functions required of such systems. The main supported aggregate operations include calculating the number of events that have arrived, the number of different events that have occurred, the sum of the values stored in them, the average, maximum and minimum. Thanks to their continuous calculation, it is possible to monitor various indicators of business processes online and quickly respond to changes in them;
- Window semantics. Calculating the results of aggregate operations on all events in the stream from the moment the request is set in the system until it is removed is not the most important task in systems designed for continuous recalculation of various metrics. A more important property is the support of advanced window semantics—the ability to group events in a stream into so-called "windows" and calculate the result of executing continuous queries over them. Examples of such queries can be: counting the number of different events in the last minute—the query result should be updated every time a new event arrives, and it is also necessary to take into account the age of events, after one minute they should be excluded from the result; output the average price of transactions for every ten minutes—in this case, a new result will appear every ten minutes and take into account only those events that occurred during them.

Basically, there are two parameters that can characterize the window semantics in a continuous query. The first is a way to set the window size. It is usually measured either by the number of events that have arrived (the window includes the last N events), or by their age (events that have arrived in the last N seconds/minutes/...). The second is the "behavior" of the window. There are three main ways: either the windows, in fact, are not used and the entire stream is one large window, or the window "slides" along the stream to the right with the arrival of new events, and the oldest ones are excluded from the window ("sliding window"). Or the window "steps" along the stream—after the window is fully processed, all events from it are discarded and the window is filled with new events, so each event will be counted only once (the "walking" window).

Another important aspect for continuous query execution systems is the possibility of its integration with as many different systems as possible, based on the event streams of which continuous queries can be performed. They can generate events in different formats, which may contain different sets of information with different structures. Therefore, it is important not to limit the format of incoming events to strict limits, but to use some common weakly structured format, for example, JSON [5]. Unlike another common weakly structured XML format [6], JSON is not redundant and provides a fairly compact record for the data stored in it. And in streams with a high frequency of events, it is important that the format of their recording is as little redundant as possible, so as not to generate a lot of useless data. In addition, JSON is currently the standard for receiving information from many web services, in particular, responses to API calls are provided in this format. Therefore, using JSON as an incoming event format for a continuous request processing system is a justified choice.

One of the general trends in modern high-load applications and web services is the rapid growth of the amount of data that you have to work with. Since the 2000s, it became clear that it is impossible to process them on one computer. Therefore, the use of complex computing technologies was required. Moreover, the volumes of data to be processed grew at a huge pace, and one of the important requirements for such technologies was to support horizontal scalability—the ability to increase system performance by adding new resources, rather than improving the characteristics of old ones. In practice, this is expressed in the fact that when using supercomputers, we may at some point run into the "ceiling" of its performance, and then we will have to change the entire expensive machine entirely. If you use a cluster of regular machines, you can simply add new nodes to improve its performance. From an economic point of view, it is preferable to use conventional non-specialized equipment available for open sale (commodity hardware). In this regard, the main architecture for building complex systems for processing large amounts of data in highly loaded web applications has become a cluster of homogeneous machines that do not have shared resources (shared-nothing) connected to each other by a conventional network, for example, Ethernet. In the conditions of such an architecture, there is no special support from the hardware for distributed computing, and therefore, all the care for organizing and managing parallelization falls on the software part. This caused the rapid development of the field of distributed computing, as a result of which a large stack of open-source software was developed, which makes it easy to develop programs that require scalable and fault-tolerant distributed computing. The pioneers in this field can be considered Google, within which such widespread technologies as MapReduce [7], GFS [8], and many others were developed. Big data also appears in streaming processing tasks, for example, the frequency of arrival of events in some threads can reach tens, and sometimes hundreds of thousands of events per second. And for such tasks, there are requirements for parallelism, scalability and fault tolerance of systems that solve them. Consequently, the use of technologies for distributed computing was required. In this regard, in the last few years, solutions have begun to appear that are specialized for the organization of distributed streaming computing, such as S4 [9] and Storm [10]. But they are still quite young, and a system designed to perform continuous queries over streaming data has not yet been developed on the basis of them. The purpose of this work is to analyze a distributed system for performing continuous queries on JSON streaming data based on modern technologies for streaming computing, suitable for solving analytics problems and calculating metrics in real time.

The task of this work is to analyze a distributed system for executing continuous queries over JSON stream data based on modern technologies for parallel stream computing. The following main subtasks can be distinguished: overview of existing systems for performing continuous queries on streaming data; research of modern approaches to the construction of parallel streaming data processing systems.

3   Results and Discussion

We will discuss the systems, their functionality and architecture, as well as highlight the general scheme of building such systems, consider the pros and cons of various approaches.

The field of streaming data processing has existed for quite a long time, and active academic research in the field of continuous query execution systems and the development of such systems using the obtained scientific results has been conducted since about the end of the 1990s. Of course, during this time, computer technologies have developed greatly, and the ideas about which platforms such systems should be built for, as well as the requirements for their performance, have changed significantly. But nevertheless, it is possible to distinguish a number of aspects that sufficiently characterize the system of performing continuous queries, regardless of its modernity. We will use them when considering existing solutions:

– Format of incoming events. Despite the fact that the work is focused on weakly structured data in JSON format, many existing systems are designed to work with event streams with a special and sometimes strictly defined structure, which, nevertheless, does not make them uninteresting for consideration;
– The purpose of the system. Streaming data processing systems are built for different purposes—some tasks include only filtering suitable events at the maximum possible speed, while others require a complex recalculation of various statistics, sometimes requiring access to external data sources. Depending on the purpose, the architectures of the systems and the methods of processing requests in them are radically different;
– Query language. Another parameter that can be used to characterize the system is the language itself for setting continuous queries. The more expressive it is, the more complex queries the system will be able to process. Many languages for setting continuous queries are based on DBMS query languages, for example, SQL;
– The method of processing the event stream. Many continuous query execution systems are based on other solutions that were not originally intended for streaming processing, for example, database management systems. They may not support the streaming data processing model at all, but only constantly recalculate the result of a continuous query as a whole, but, nevertheless, they cope well with the task. To develop your own system, it is important to understand the existing approaches, their advantages and disadvantages.

3.1 RDBMS-Based Systems

The idea that systems for performing continuous queries over streaming data can be built on the basis of relational database management systems is quite logical—incoming events can be represented as strictly typed tuples, similar to those stored in a database, and the functionality in the languages of continuous queries and database query languages largely intersects. For example, consider the SQL language: the functionality provided by the main select, where, group by operators, which are responsible, respectively, for selecting, filtering and grouping data, is necessary in the same form and when executing continuous queries. The exception is that in the case of a database management system, all incoming tuples are already known, and you can immediately execute a query on all the data, and in the case of a continuous query, you need to wait for new tuples to arrive and work with them as they appear. Another important difference between RDBMS and systems for executing continuous queries based on them is that some SQL statements, such as order by, join, cannot be used unchanged to execute continuous queries—their result is determined only in cases when the full set of input data is known, which is impossible with the constant arrival of new events. But their functionality is also in demand, so support for window semantics is required. For example, all the events received during these five minutes can be fed to the input of the order by operator every five minutes-and then we will periodically receive a sorted result. Note that to support window semantics in the query language originally developed for the DBMS, it is necessary to add special constructs. As for the architectural part, the following RDBMS components can be reused with minor changes when building continuous query execution systems:

– Query analyzer. Responsible for parsing the request in accordance with the query language and building a plan for its execution. The changes consist in expanding the query language with window semantics;
– Query optimizer. The parts concerning optimization of access methods to data stored on disk should be excluded from the classical cost optimizer, since such operations are no longer used when performing continuous queries;
– Query executor. The classic query executor that implements traversal according to the query execution plan may remain almost unchanged, the changes will consist in the implementation of operations that require window semantics. Also, some RDBMS components are not necessary in continuous query execution systems;
– Data storage subsystem. In a purely streaming data processing model, saving incoming events on disk is not required at all, although it can be quite a useful feature;
– Index subsystem;
– Support for transactions, logging, and crash recovery. Usually, the purpose of systems for executing continuous queries based on RDBMS is to process events from many sources that can be analyzed together (thanks to the connection mechanism inherited from RDBMS). In particular, a prominent representative of such systems is Tele-graphCQ [11], developed by a group of databases at Berkley University based on the PostgreSQL DBMS. Another example is the JSQ system for performing continuous queries over data streams in JSON format, based on the AmosQL DBMS [12]. Its peculiarity is that incoming events must be stored in the DBMS before processing.

3.2   XML Streaming Processing Systems

A large subdomain in the streaming processing and execution of continuous queries is the processing of events in the XML format [6], which allows storing weakly structured data. The popularity of this area is associated with the convenience and prevalence of XML as a format for the interaction of various applications with each other. Also, thanks to the XML structure, there is a great potential for developing query languages and efficient models for their execution. Systems for performing continuous queries over XML began to develop in the late 1990s, and are developing to this day. The general principle of executing continuous queries does not differ from that in RDBMS-based systems: first you need to build a query execution plan, optimize it if possible, and then pass it to the executor, who will receive events and calculate the result. The differences are that now incoming events are not records with fields of a fixed type, but objects with a potentially complex and varying hierarchical structure that you need to work with. A large number of continuous query languages for XML have also been developed, with varying degrees of expressiveness. In particular, tools for setting continuous queries are included in the latest, at the moment, XQuery language standard [13]. One of the interesting systems for performing continuous queries over event streams in XML format is NiagaraCQ [14]. Its main purpose is to process a large number of similar requests (up to several thousand) to exchange indicators. XML-QL is used as the query language [15]. It is assumed that the system works with a large number of users who set their queries to a single information stream. In this case, the flow of incoming events does not have to be large in order to cause performance problems, because you need to serve a large number of requests. However, since the incoming event stream may not contain a very large number of different entities (stock symbols), it is very likely that the requests will overlap strongly and there is no need to calculate the same values several times. The NiagaraCQ system implements an advanced technique for identifying common parts in queries and grouping them for joint execution. When analyzing a query, its signature is built, which consists in identifying structural predicates, and removing constant symbols. Requests with the same signature form a group of requests, which, when a new event arrives, can be executed only once, and not for each request individually. Each group has one query execution plan and a constant table containing records about each query of the group. The query in the group is characterized by the requested values for comparison and the delivery address of the result. Groups are formed as requests are received. When a new one is added, its signature begins to be compared from the bottom up with the signatures of already existing query groups. If at one point it turns out that the signature of one of the groups matches the scanned part of the request signature, then the request is divided into two—the one whose signature matches the found group, and the remaining one. The same search is performed for the remaining query. If it is possible to find groups for all parts of the query, its execution will have almost no effect on the system performance, since no new calculations were added. If there were no suitable groups for some of the parts, new ones will be created.

Also, the problem of splitting existing groups into subgroups is solved, in order to optimize their number and increase productivity accordingly. Another important task solved by continuous query execution systems is to identify the sequences of events of interest. For example, you may need to track those stock symbols whose price has risen and then fallen over the past day. This requires additional support from the query language, and serious modifications in the query executor. In particular, the XSeq language [16] was developed, which is an extension of the XPath language [17] to support the detection of event sequences. One of the ways to build a continuous query execution plan for XML is to convert the query to a finite state machine. The methods of constructing automata vary depending on the required functionality. In particular, the authors of the XSeq language offer a similar effective implementation of their language.

3.3 Specialized Streaming Processing Systems

Many continuous query execution systems were built to solve specific tasks, and therefore their functionality may be limited to the scope of the task needs. But nevertheless, they are also of interest. One example of a specialized system for executing continuous requests is the Gigascope system [18], developed for monitoring the state of networks and traffic in them at AT&T. The list of tasks it solves includes monitoring network bandwidth, monitoring the status of network equipment, detecting suspicious user activity on the network. Incoming events are packets of network protocols of various levels coming from the network and restored by special utilities. A specially developed GSQL language is used as the query language—a narrowing of the SQL language, expanded by window semantics. The main emphasis in the system is on the efficient execution of queries that require window semantics. For their effective implementation, non-blocking implementations of connection and aggregation operators are proposed. The internal structure of the Gigascope system is also of interest. The scheme of parsing and building a continuous query plan remains similar to the systems discussed earlier, but the principle of operation of the query executor differs. Gigascope is a program code generator—according to the built tree (plan) of query execution, for each individual operation, its own program module is generated in C or C++ with a specific interface. After generating all the necessary code, it is compiled and compiled with a special module called the "thread manager", which is responsible for receiving incoming events and transmitting them to the generated modules responsible for performing individual operations. As a result, setting a continuous query leads to the generation of a separate program for its execution. According to the authors of Gigascope, this approach is the most productive, which is quite understandable, because the module for each operation is optimal for this request, and does not contain redundant language constructs and checks for those cases that do not occur in the current continuous request. Another interesting example of a specialized system for executing continuous requests is the SASE system [19], designed to analyze message flows coming from RFID tag scanners [1]. The proposed tasks for solving with the help of SASE are the detection of shoplifting, automation of supply management systems, automation of inventory control, etc. The main required functionality when analyzing event flows of this kind is the ability to identify complex predefined sequences of events in real time that correspond to the passage of observed objects through different scanners, and transmit this information to other systems for further processing. For the SASE system, a continuous query execution language of the same name is proposed, which supports only the detection of sequences of events. A nondeterministic finite automaton is built based on a given query, which allows you to identify interesting sequences of events at high speed.

3.4   Non-threaded Processing of Continuous Requests

To perform continuous queries, a purely streaming processing model is not necessarily required—when the result of executing a query is not recalculated entirely, but only some part of it is updated when new events are received. Many systems perform continuous queries by completely recalculating the query result. Let's consider one of them. One of the frequent applications of continuous query execution systems is the calculation of various metrics of observed processes in real time. Examples include calculating the average transaction price for the last day in an online store, calculating the percentage of requests that were answered above a specified time threshold when monitoring the stability of the web service, and many others. The Cube system [20] is designed to calculate such metrics. It accepts events in a weakly structured JSON format [5] and executes continuous queries on them, specified in a special language. The main functionality is to support the calculation of aggregate values (sum, average, maximum/minimum, etc.) over the data stored in the received events. Window semantics is supported in the form of "walking" windows—metric values can be recalculated over parts of the stream only at certain intervals. Filtering of events by various predicates is also supported. Each event that comes to the Cube system is saved to a database stored in the MongoDB DBMS [21]. If it is necessary to update the result of a continuous query (the time interval after which a new result is required to be issued has ended), a distributed MapReduce task [7] is launched over the incoming events, which recalculates the result. Due to the fact that the MongoDB DBMS is distributed and supports the execution of MapReduce tasks on stored data, acceptable performance and scalability of the solution is achieved [22–25]. To speed up the execution of continuous queries, their results are cached for each query and can be reused when calculating other queries.

4 Conclusions

To implement the tasks of analyzing large amounts of data, there is a problem associated with the need to use high-performance equipment and computer technologies that allow processing such huge amounts of information. In fact, big data is presented as a kind of obstacle (this is a new form of information barrier). On the one hand, we see that big data will allow humanity to solve new information and technological problems. On the other hand, their development determines the solution of promising projects related to the development of complex and integrated systems and technologies. The existing systems for executing continuous queries and their architecture were considered above. We can distinguish a general scheme of such systems. In any of them, due to the presence of a high-level query language, the execution of a query begins with the construction of a plan for its execution. The query execution is handled by the executor, which is most often a classic hierarchical query executor, similar to that in database management systems. The currently existing systems for executing continuous queries over data streams in JSON format do not meet all the requirements imposed on them, since they are not pure streaming processing systems. They are either based on some database management system and require mandatory saving of events in it, which can significantly slow down processing, or they do not provide a streaming model for executing continuous queries, but only periodically recalculate its result, which contradicts the requirement for the fastest processing of incoming events. Thus, it seems appropriate to develop a proprietary system for executing continuous queries over JSON streaming data based on modern technologies for streaming computing.

References

1. Denisova, O.A., Kunsbaeva, G.A., Chiglintsiva, A.S.: Big data: some ways to solve the problems of higher education. In: Journal of Physics: Conference Series. International Scientific and Practical Conference Information Technologies and Intelligent Decision Making Systems, ITIDMS-II 2021, p. 012021 (2021)
2. Denisova, O.A.: Motivation of technical university students to study physics and methods of teaching it in the context of a pandemic. In: Journal of Physics: Conference Series, p. 12025. Krasnoyarsk Science and Technology City Hall (2020)
3. Denisova, O.A.: Big data technology: assessing the quality of the educational environment. In: Journal of Physics: Conference Series. International Scientific and Practical Conference Information Technologies and Intelligent Decision Making Systems, ITIDMS-II 2021, p. 012027 (2021)

4. Sharma, S., Gadia, S., Udoyara, S.: Subset, subquery and queryable-visualization in parametric big data model. Int. J. Inf. Manag. Data Insights **1**(1), 100003 (2021)
5. Brahmia, Z., Grandi, F., Brahmia, S.: A graphical conceptual model for conventional and time-varying JSON data. Procedia Comput. Sci. **184**, 823–828 (2021)
6. XML: eXtensible Markup Language. http://www.w3.org/XML/. Accessed 21 Oct 2021
7. Dean, J., Ghemawat, S.: MapReduce: simplified data processing on large clusters. In: OSDI 2004: Proceedings of the 6th Conference on Symposium on Operation Systems Design and Implementation. USENIX Association (2004)
8. Ghemawat, S., Gobioff, H., Leung, S.: The Google file system. SIGOPS Oper. Syst. Rev. **37**(5), 29–43 (2003)
9. Neumeyer, L., Robbins, B., Nair, A., Kesari, A.: S4: distributed stream computing platform (2010)
10. Wang, J., Zhang, J., Zhong, R.: Big data analytics for intelligent manufacturing systems. J. Manuf. Syst. 1016 (2021)
11. Chandrasekaran, S., Cooper, O.: TelegraphCQ: continuous dataflow processing for an uncertain world. In: CIDR, vol. 20, p. 668 (2003)
12. Gray, P., Amosql, M.: Liu, L., Ozsu, M.T. (eds.) Encyclopedia of Database Systems. Springer (2019)
13. XQuery: An XML Query Language. http://www.w3.org/TR/xquery-30/. Accessed 21 Oct 2021
14. Chen, J., DeWitt, D.J., Tian, F., Wang, Y.: NiagaraCQ: a scalable continuous query system for internet databases. In: SIGMOD 2000, pp. 379–390 (2000)
15. XML-QL: A Query Language for XML. http://www.w3.org/TR/NOTE-xml-ql/. Accessed 21 Oct 2021
16. Wu, E., Diao, Y., Rizvi, S.: High-performance complex event processing over streams. In: Proceedings of the 2006 ACM SIGMOD International Conference on Management of Data, pp. 407–418 (2006)
17. XPath: XML Path Language. http://www.w3.org/TR/xpath20/. Accessed 21 Oct 2021
18. Cranor, C., Johnson, T., Spataschek, O., Shkapenyuk, V.: Gigascope: a stream database for network applications. Network, pp. 647–651 (2013)
19. Gyllstrom, D., Sase, Wu, E.: Complex event processing over streams. In: Proceedings of the Third Biennial Conference on Innovative Data Systems Research (2007)
20. Cube: A system for collecting timestamped events and deriving metrics. https://github.com/square/cube
21. Mongo, D.B.: An open-source document database. http://www.mongodb.org/. Accessed 21 Oct 2021
22. Benítez-Hidalgo, A.: TITAN: a knowledge-based platform for Big Data workflow management. Knowl.-Based Syst. **232**, 107489 (2021)
23. Bauleo, E., Carnevale, S.: Design, realization and user evaluation of the SmartVortex Visual Query System for accessing data streams in industrial engineering applications. J. Vis. Lang. Comput. **25**(5), 577–601 (2014)
24. Sahal, R., Breslin, J.G., Intizar, M.: Big data and stream processing platforms for Industry 4.0 requirements mapping for a predictive maintenance use case. J. Manuf. Syst. **54**, 138–151 (2020)
25. Chu, Z., Yu, J., Hamdulla, A.: A novel deep learning method for query task execution time prediction in graph database. Future Gener. Comput. Syst. **112**, 534–548 (2020)

Implementation of Sustainability Analyzes in Software Products for Evaluating the Effectiveness of Investment Projects

Kirill Zhichkin[1]($\boxtimes$) (ID), Vladimir Nosov[2,3] (ID), Aleksandr Zhichkin[4] (ID), and Aleksandra Łakomiak[5] (ID)

[1] Samara State Agrarian University, Kinel, Russian Federation
zskirill@mail.ru

[2] Kutafin Moscow State Law University, Moscow, Russian Federation

[3] Academy of the Investigative Committee of the Russian Federation, Moscow, Russian Federation

[4] National Research Nuclear University MEPHI, Moscow, Russian Federation

[5] Finance University under the Government of the Russian Federation, Moscow, Russian Federation

Abstract. Analysis of the sustainability of an investment project is the most important part of assessing its effectiveness. Modern methods of stability analysis make it possible not only to assess its results in the event of the most probable scenario, but also to determine the boundaries of the permissible change in the factors of the external and internal environment. The purpose of the study is to identify ways to improve the methodology for conducting sustainability analyzes in software products used to assess the effectiveness of investment projects. Within the framework of this study, the following tasks are expected to be solved: - to consider the methods of conducting stability analyzes used in various software products; - to identify the shortcomings of the methods used; - to propose ways to improve the implementation of methods for analyzing sustainability in the considered software products used to assess the effectiveness of investment projects. As a result of the study, possible directions for improving the most common software products used to assess the effectiveness of investment projects were identified. An introduction to the analysis of two-factor sensitivity analysis, limiting the period of coverage of the break-even analysis and an algorithm for automating the statistical analysis is proposed.

Keywords: Sustainability analysis · Sensitivity analysis · Statistical analysis · Break-even analysis · Software · Investment project · Monte Carlo analysis

1 Introduction

Determining the effectiveness of an investment project is an important, multi-stage task that includes calculations for all stages of the pre-investment, investment and production phases of investment activities [1–5]. Accurate display of all parameters and features of

© The Author(s), under exclusive license to Springer Nature Switzerland AG 2022
A. Gibadullin (Ed.): DITEM 2021, LNNS 432, pp. 59–69, 2022.
https://doi.org/10.1007/978-3-030-97730-6_6

each stage allows you to predict the results of the project with high accuracy, determine all its strengths and weaknesses, determine in advance the time intervals that require increased attention of the project operator, and choose a strategy for its implementation [6–12].

The main tool for evaluating an investment project is numerous indicators characterizing certain aspects of investment, production, financial, etc. activities. The modern methodology for evaluating an investment project provides information on the current and forecast state of the project parameters and, at the same time, offers a set of standard methods for correcting each problem situation [13–16].

Along with performance indicators, there is a set of methods called project sustainability analysis methods. These include: sensitivity analysis (characterizing the overall stability of the project), break-even analysis (determines the parameters of the sustainability of production activities), three-component analysis (financial stability analysis) and Monte Carlo analysis (statistical analysis of the project) [17–19].

To speed up the calculations for the assessment of investment activities, the corresponding software is currently widely used. In the conditions of the Russian Federation, numerous programs of the corresponding profile are used, both developed by companies and by individual users. Among all this variety, the most widespread are two software products - Project Expert («Expert Systems» company) and Alt-Invest (Sum) («Alt-Invest» company) [20–22].

2 Materials and Methods

The purpose of the study is to identify ways to improve the methodology for conducting sustainability analyzes in software products used to assess the effectiveness of investment projects. Within the framework of this study, the following tasks are expected to be solved: - to consider the methods of conducting stability analyzes used in various software products; - to identify the shortcomings of the methods used; - to propose ways to improve the implementation of methods for analyzing sustainability in the considered software products used to assess the effectiveness of investment projects.

The break-even analysis shows how changes in individual factors of the external and internal environment affect the results of the project. If you set the interval for changing the factor, then during the calculations it is clear in what interval of its change you can get a positive result [23–28]. And the wider this interval, the more sustainable the project. There is a single-factor sensitivity analysis (when only one parameter of the project changes), which is implemented in almost all software products for assessing the effectiveness of an investment project, and a multifactor analysis - when two or more parameters are simultaneously changed [29–31].

The break-even analysis allows you to determine the volume of production at which the profitability of the project will be zero. By comparing this value with the planned production volume, you can determine the safety margin of the project. The larger the planned volume, the more sustainable the project [32–36].

Monte Carlo analysis (statistical analysis) allows you to simulate the probability of successful project implementation and calculate, taking into account its updated values of performance indicators [37–40].

3   Results

This study examines the implementation of sustainability methods in the latest versions of the most common software products - Project Expert version 7.57 and Alt-Invest version 6.0.

The Project Expert program is a closed-type software package. Its structure does not allow one to see the calculation formulas used in it, but thanks to this approach, the program has a more flexible calculation apparatus and easily adapts to changes in external environmental factors (legislation, methods for calculating indicators, etc.) The applied stability calculation apparatus is the most complete. It includes all three considered methods.

Sensitivity analysis is represented by univariate analysis (in the Project Expert program) and multivariate (in the What if analysis program). A positive point is the ability to use in What if analysis of data calculated in the Project Expert program without any processing. The results are presented in the form of a table, similar to the one that is formed at the bottom of the menu (Fig. 1). It is also possible to display the results graphically (Fig. 2).

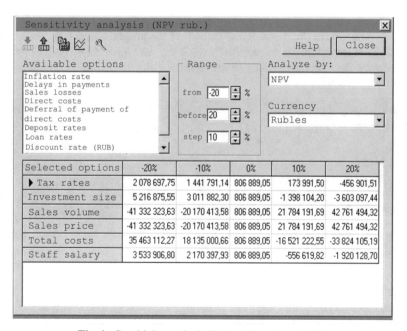

Fig. 1. Sensitivity analysis (Project Expert ver. 7.57).

The break-even analysis in the latest version of the program has significantly greater calculation capabilities compared to the previous version. If in Project Expert ver. 6, only the break-even point was calculated for the product, now you can additionally obtain data on the ratio of the break-even point and the planned production volume (for the product and the company as a whole), for the financial stability margin and for the

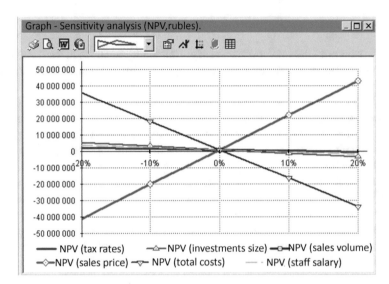

Fig. 2. Graphical display of sensitivity analysis results (Project Expert ver. 7.57).

operating leverage. In other words, the authors of the program have now gone beyond the traditional methodology of sensitivity analysis, which makes Project Expert even more informative in terms of assessing the sustainability of the project (Fig. 3).

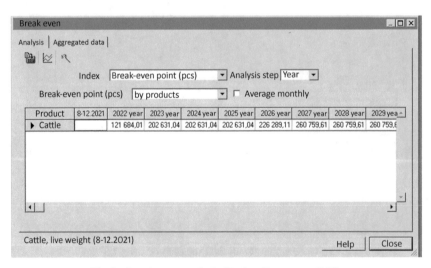

Fig. 3. Break-even analysis (Project Expert ver. 7.57).

Monte Carlo analysis allows you to simulate the likelihood of obtaining a positive result during the implementation of the project [41–43]. For this, the factors of the internal and external environment are selected that maximally affect the result, the boundaries of their change are set, and within them the values of the selected parameters are randomly

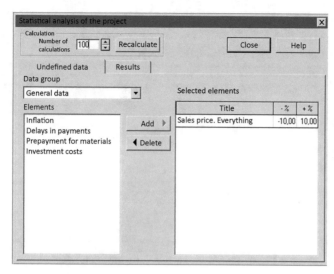

Fig. 4. Statistical analysis of the project (Project Expert ver. 7.57).

selected (Fig. 4). At least one hundred calculations are carried out and the ratio of the number of experiments with positive results to the total number of experiments performed is determined [44–46]. The higher the value obtained, the more sustainable the project is.

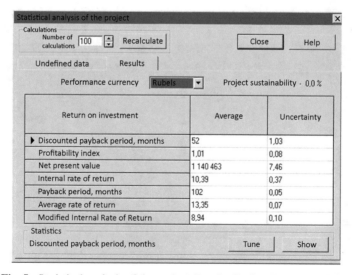

Fig. 5. Statistical analysis of the project. Results (Project Expert ver. 7.57).

2036						
2037	⋆	Sensitivity analysis		Tune		
2038						
2039	Variable parameter					
2040						
2041	Discount rate					
2042						
2043	In the interval					
2044			from	85%		
2045			with step	5%		
2046						
2047	Final indicator					
2048						
2049	NPV for total investment costs					
2050						
2051	Calculated value table			Value	Result	
2052				85%	494 166	
2053		Recalculate		90%	318 467	
2054				95%	142 295	
2055				100%	-38 700	
2056				105%	-219 695	
2057				110%	-400 690	
2058				115%	-581 685	

Fig. 6. Sensitivity analysis (Alt-Invest ver. 6).

An additional bonus of statistical analysis is the ability to calculate performance indicators taking into account the calculations performed. For this, the arithmetic mean of each indicator is calculated for all the experiments performed. It is believed that the resulting value more accurately reflects the features of the actual implementation of the project (Fig. 5).

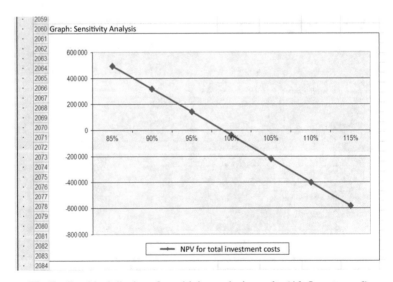

Fig. 7. Graphical display of sensitivity analysis results (Alt-Invest ver. 6).

The Alt-Invest program is an "open" type of program. Being in essence an MSExcel file with a large number of macros used, the program has received great recognition for its ease of learning, since almost all calculation formulas are presented to the user and at any time he can see how this or that indicator is calculated.

Unfortunately, the possibilities for assessing the sustainability of the project in the program are limited only by one-factor sensitivity analysis (Fig. 6) and the calculation of the break-even point in the table of financial indicators. The form of presenting the results is not the most successful, since in order to obtain comparable results for several assessed indicators, additional work is required to bring them into a single table. The same applies to the graphical form of displaying the results (Fig. 7). The graph shows the change in only one indicator, which makes it difficult to compare results and select those that have the greatest impact on the result.

4 Discussion

Comparison of methods for conducting sensitivity analysis in the Project Expert and Alt-Invest programs, which are currently the standard for assessing the effectiveness of investment projects in a large area of the world economic system, allows us to high-light the disadvantages and advantages that could be additionally implemented in these programs.

1. It is necessary to expand the capabilities of sensitivity analysis by introducing a two-factor model into the calculation. The use of sensitivity analysis by two parameters will automatically build a model of the impact on the results of two project parameters simultaneously and obtain a visual display of the data obtained in tabular and graphical form (Fig. 8).
2. When carrying out a break-even analysis, it is necessary to provide for the possibility of automatically limiting the settlement period. When determining the results of the break-even analysis, the calculation is carried out not only of those periods when the production phase begins in the investment project, but also in the pre-investment and investment phases, when there is no sense in calculating, since there is no production itself. Perhaps, an option should be provided for limiting the calculation period of the break-even analysis associated with the schedule of the project implementation - with the beginning of the "Production" stage.
3. Statistical analysis. When carrying out it, it is necessary to provide for an automatic calculation option for the most important parameters of the project. It is believed that the greatest influence on the results is exerted by those factors that, during the sensitivity analysis, form 80% of the deviations from the abscissa axis (Fig. 2). Therefore, it is necessary to introduce the following calculation algorithm. Initially, a one-way sensitivity analysis is carried out for all parameters that affect the project's outcome. Then, in automatic mode, only those are selected that have a significant impact on the result "80:20 principle. And in the conclusion, a calculation is carried out, both for each parameter separately, and a combined statistical analysis for all selected factors. By presenting the result in tabular and graphical form, it is possible to determine both the influence of individual factors and the general one on the sustainability of the project.

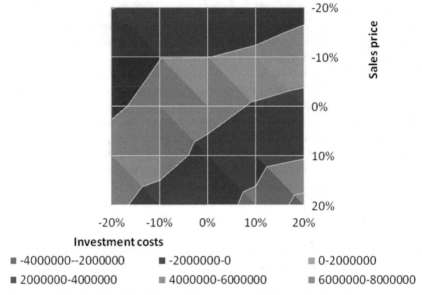

Fig. 8. Graphical display of two-way sensitivity analysis results.

4. For the Alt-Invest program, it is proposed to gradually expand the capabilities of stability analyzes in the following areas: the formation of a combined one-factor sensitivity analysis in tabular and graphical form for the possibility of comparing the results with each other; introduction of a two-factor sensitivity analysis model into the calculation; formation of a break-even analysis, similar to that available in the Project Expert program.

5   Conclusion

Currently, in Russia, the Baltic countries and the CIS, programs for evaluating the effectiveness of investment projects Project Expert and Alt-Invest are widely used. They implement various approaches to the formation of these investment projects. An important part of the assessment is to conduct stability analyzes in relation to changes in external and internal environmental factors. As the analysis of the above programs shows, it is necessary to improve the existing models implemented in these software products. Among the main improvements, it is necessary to highlight: the introduction of two-factor sensitivity analysis; limiting the period for calculating the results of the break-even analysis; automation of statistical analysis taking into account sensitivity analysis data.

References

1. Gupta, M., Kohli, A.: Enterprise resource planning systems and its implications for operations function. Technovation **26**(5–6), 687–696 (2006)
2. Zhichkin, K.A., Nosov, V.V., Zhichkina, L.N., Pavlyukova, A.V., Korobova, L.N.: Modeling the production activity of personal subsidiary plots in the regional food security system. In: IOP Conference Series: Earth and Environmental Science, vol. 659, p. 012005 (2021)
3. Stone, R.A.: Leadership and information system management: a literature review. Comput. Hum. Behav. **10**(4), 559–568 (1994)
4. Gholamzadeh Chofreh, A., Goni, F.A., Ismail, S., Mohamed Shaharoun, A., Klemeš, J.J., Zeinalnezhad, M.: A master plan for the implementation of sustainable enterprise resource planning systems (part I): concept and methodology. J. Clean. Prod. Part B **136**, 176–182 (2016)
5. Susanto, A., Meiryani, M.: The impact of environmental accounting information system alignment on firm performance and environmental performance: a case of small and medium enterprises s of Indonesia. Int. J. Energy Econ. Policy **9**(2), 229–236 (2019)
6. Tryhuba, A., et al.: Forecasting quantitative risk indicators of investors in projects of biohydrogen production from agricultural raw materials. Processes **9**(2), 258 (2021)
7. Wieder, B., Booth, P., Matolcsy, Z.P., Ossimitz, M.-L.: The impact of ERP systems on firm and business process performance. J. Enterp. Inf. Manag. **19**(1), 13–29 (2006)
8. Deptuła, A.M., Knosala, R.: Risk assessment of the innovative projects implementation. Manag. Prod. Eng. Rev. **6**(4), 15–25 (2015)
9. Madanhire, I., Mbohwa, C.: Enterprise resource planning (ERP) in improving operational efficiency: case study. Procedia CIRP **40**, 225–229 (2016)
10. Sborshikov, S., Vvedenskiy, R., Markova, I.: The application of simulation modelling in making operational decisions in construction. In: IOP Conference Series: Materials Science and Engineering, vol. 1030, p. 012106 (2021)
11. Kitchenham, B., et al.: Systematic literature reviews in software engineering-a tertiary study. Inf. Softw. Technol. **52**(8), 792–805 (2010)
12. Zhichkin, K., Nosov, V., Zhichkina, L., Panchenko, V., Zueva, E., Vorob'eva, D.: Modelling of state support for biodiesel production. In: E3S Web of Conferences vol. 203, p. 05022 (2020)
13. Shadroo, S., Rahmani, A.M.: Systematic survey of big data and data mining in Internet of Things. Comput. Netw. **139**, 19–47 (2018)
14. Spathis, C., Constantinides, S.: The usefulness of ERP systems for effective management. Ind. Manag. Data Syst. **103**(8–9), 677–685 (2003)
15. Patrício, D.I., Rieder, R.: Computer vision and artificial intelligence in precision agriculture for grain crops: a systematic review. Comput. Electron. Agric. **153**, 69–81 (2018)
16. Zhichkin, K.A., Nosov, V.V., Zhichkina, L.N., Ramazanov, I.A., Kotyazhov, A.V., Abdulragimov, I.A.: The food security concept as the state support basis for agriculture. Agron. Res. **19**(2), 629–637 (2021)
17. Abar, S., Theodoropoulos, G.K., Lemarinier, P., O'Hare, G.M.P.: Agent based modelling and simulation tools: a review of the state-of-art software. Comput. Sci. Rev. **24**, 13–33 (2017)
18. Danylyshyn, B., Bondarenko, S., Malanchuk, M., Kucherenko, K., Pylypiv, V., Usachenko, O.: Method of real options in managing investment projects. Int. J. Innov. Technol. Explor. Eng. **8**(10), 2696–2699 (2019)
19. Çizakça, M.: Risk sharing and risk shifting: an historical perspective. Borsa Istanbul Rev. **14**(4), 191–195 (2014)
20. Riepina, I., Hrybinenko, O., Parieva, N., Parieva, O., Savenko, I., Durbalova, N.: Quantity assessment of the risk of investment projects. Int. J. Recent Technol. Eng. **8**(3), 7256–7260 (2019)

21. Zhichkin, K., Nosov, V., Zhichkina, L.: The express method for assessing the degraded lands reclamation costs. In: Mottaeva, A. (ed.) Proceedings of the XIII International Scientific Conference on Architecture and Construction 2020: Commemorating the 90th anniversary of Novosibirsk State University of Architecture and Civil Engineering, pp. 483–492. Springer, Singapore (2021). https://doi.org/10.1007/978-981-33-6208-6_47
22. Gibadullin, A.A., Yurieva, A.A., Morkovkin, D.E., Isaichykova, N.I.: Monitoring the reliability and efficiency of the electricity industry. In: IOP Conference Series: Materials Science and Engineering, vol. 919, no. 6, p. 062018 (2020)
23. Zhichkin, K., Nosov, V., Zhichkina, L., Aydinov, H., Arefiev, I., Kuznetsova, I.: Formalization of risk analysis in software products for calculating the effectiveness of investment projects. J. Phys. Conf. Ser. **2001**, 012016 (2021)
24. Kamilaris, A., Kartakoullis, A., Prenafeta-Boldú, F.X.: A review on the practice of big data analysis in agriculture. Comput. Electron. Agric. **143**, 23–37 (2017)
25. Chofreh, A.G., Goni, F.A., Klemeš, J.J., Malik, M.N., Khan, H.H.: Development of guidelines for the implementation of sustainable enterprise resource planning systems. J. Clean. Prod. **244**, 118655 (2020)
26. Morkovkin, D.E., Kolosova, E.V., Sadriddinov, M.I., Semkina, N.S., Gibadullin, A.A.: Organizational and management mechanisms for the digital transformation of economic activities. In: IOP Conference Series: Earth and Environmental Science, vol. 507, p. 012023 (2020)
27. Zhichkin, K.A., Starikov, P.V., Zhichkina, L.N., Mamaev, O.A., Artemova, E.I., Levochkina, N.A.: The applied software role in the training of economic specialties students. J. Phys. Conf. Ser. **1691**, 012111 (2020)
28. Tryhuba, À., Boyarchuk, V., Tryhuba, I., Ftoma, O., Padyuka, R., Rudynets, M.: Forecasting the risk of the resource demand for dairy farms basing on machine learning. In: CEUR Workshop Proceedings, vol. 2631, pp. 327–340 (2020)
29. Angelopoulos, D., Doukas, H., Psarras, J., Stamtsis, G.: Risk-based analysis and policy implications for renewable energy investments in Greece. Energy Policy **105**, 512–523 (2017)
30. Romanova, Ju.A., Morkovkin, D.E., Romanova, Ir.N., Artamonova, K.A., Gibadullin, A.A.: Formation of a digital agricultural development system. In: IOP Conference Series: Earth and Environmental Science, vol. 548, no. 3, p. 032014 (2020)
31. Li, C.-B., Lu, G.-S., Wu, S.: The investment risk analysis of wind power project in China. Renew. Energy **50**, 481–487 (2013)
32. Zhichkin, K., Nosov, V., Zhichkina, L., Abdulragimov, I., Kozlovskikh, L.: Formation of a database on agricultural machinery for modeling the production cost. In: CEUR Workshop Proceedings, vol. 2922, pp. 155-163 (2021)
33. Bottou, L., Curtis, F.E., Nocedal, J.: Optimization methods for large-scale machine learning. SIAM Rev. **60**(2), 223–311 (2018)
34. Tryhuba, A., et al.: Risk assessment of investments in projects of production of raw materials for bioethanol. Processes **9**(1), 12 (2020)
35. Gadanakis, Y., Bennett, R., Park, J., Areal, F.J.: Evaluating the sustainable intensification of arable farms. J. Environ. Manag. **150**, 288–298 (2015)
36. Zhichkin, K., Nosov, V., Zhichkina, L.: The production costs calculation automation for planning the crops production parameters. In: CEUR Workshop Proceedings, vol. 2843, p. 20 (2021)
37. Kim, Y.-J.: Monte Carlo vs. fuzzy Monte Carlo simulation for uncertainty and global sensitivity analysis. Sustainability **9**(4), 539 (2017)
38. Aloini, D., Dulmin, R., Mininno, V.: Modelling and assessing ERP project risks: A Petri Net approach. Eur. J. Oper. Res. **220**(2), 484–495 (2012)
39. Zhichkin, K., Nosov, V., Zhichkina, L., Fomenko, N.: Simulation modeling in assessing the agricultural enterprise state in an emergency. In: E3S Web of Conferences, vol. 285, p. 01010 (2021)

40. Mitropoulos, L.K., Prevedouros, P.D., Yu, X., Nathanail, E.G.: A Fuzzy and a Monte Carlo simulation approach to assess sustainability and rank vehicles in urban environment. Transp. Res. Procedia **24**, 296–303 (2017)
41. Zhou, J., Reniers, G.: Modeling and application of risk assessment considering veto factors using fuzzy Petri nets. J. Loss Prev. Process Ind. **67**, 104216 (2020)
42. Neelakanta, P.S., De Groff, D.F.: Neural Network Modeling: Statistical Mechanics and Cybernetic Perspectives, pp. 1–240 (2018)
43. Sadeghi, N., Fayek, A.R., Pedrycz, W.: Fuzzy Monte Carlo simulation and risk assessment in construction. Comput.-Aided Civ. Infrastruct. Eng. **25**(4), 238–252 (2010)
44. Hellel, E.K., Hamaci, S., Ziani, R.: Modelling and reliability analysis of multi-source renewable energy systems using deterministic and stochastic Petri net. Open Autom. Control Syst. J. **10**, 25–40 (2018)
45. Dheskali, E., Koutinas, A.A., Kookos, I.K.: Risk assessment modeling of bio-based chemicals economics based on Monte-Carlo simulations. Chem. Eng. Res. Des. **163**, 273–280 (2020)
46. Liu, L., Liu, X., Liu, G.: The risk management of perishable supply chain based on coloured Petri Net modeling. Inf. Process. Agric. **5**(1), 47–59 (2018)

Support of Design Decision-Making Process Using Virtual Modeling Technology

V. Nemtinov[1]([⊠]) [iD], S. Egorov[1] [iD], A. Borisenko[1] [iD], V. Morozov[1] [iD], and Yu. Nemtinova[2] [iD]

[1] Department of Computer-Integrated Systems in Mechanical Engineering, Tambov State Technical University, 106, Sovetskaya Street, Tambov 392000, Russian Federation
nemtinov.va@yandex.ru
[2] Department of Management, Marketing and Advertising, Tambov State University named after G R Derzhavin, 33, Internatsionalnaya Street, Tambov 392000, Russian Federation

Abstract. The article considers the issues of information support for decision-making in industrial design by using virtual modeling technology. The authors have developed an information-logical model for technological equipment placement, which allows to find optimal options for technological equipment layout in an automated mode that ensures compliance with the norms and rules of industrial facilities layout. The implementation of the model was carried out in a virtual environment created by using: the SOLIDWORKS graphics system, the Twinmotion program, which provides immersive architectural 3D visualization, as well as 3DVista Virtual Tour Suite for creating virtual tours of the projected industrial facility.

Keywords: Digital transformation of production · Virtual modeling · Design solutions

1 Introduction

Digitalization of society is now the main task of the economy, which in turn is impossible without digitalization of industry. Virtual and augmented reality tools can be actively used throughout the entire product lifecycle: to identify errors at the earliest stages of design before passing mock-up commissions, to improve ergonomics and the entire production process as a whole, modelling the processes of operation, modernization and repair [1].

VR/AR technologies allow to be at the operator's workplace inside the designed product, check its various operational characteristics, technical requirements, access to various components of the product for the convenience of installation and repair. The scope of application of VR/AR technologies in production is actively developing to simulate the installation of new equipment, which saves time spent on its design. The developed equipment models can be used simultaneously to train operators working on them. As a result, employees at the new site will be immediately ready to commence work, which saves time on training.

A. Gibadullin (Ed.): DITEM 2021, LNNS 432, pp. 70–77, 2022.
https://doi.org/10.1007/978-3-030-97730-6_7

Such parallel processes make it possible to speed up production lines and equipment setup, which will have a positive economic effect for the enterprise [2]. Virtual and augmented reality technologies allow for more efficient equipment layout and to choose the optimal color solutions for the designed products that meet the customer's needs. With the help of VR/AR technologies, it is possible to simulate an emergency situation. It is often rather dangerous and expensive to model emergencies in reality, so the staff is only familiar with the theory of how to act in such cases. Virtual reality allows the company's employees to be more prepared for action in any dangerous situation. Digital space allows to carry out design and discussion, simultaneously connecting specialists of various fields in a single information field, so that they do not need to be in the same room, but to whom it is important to be in the same virtual space, which is provided by VR helmets and special software. All this allows to save time on business trips and projects coordination.

At the moment, almost all large production facilities are switching to the PLM-geometric model as the main element of the virtual prototyping approach (Product Life-cycle Management). CAD packages have already been used for product design for quite a long time, and in many cases the work is carried out with a 3D model of products. Virtual prototyping is impossible without it. The described model is usually referred to as a digital prototype or digital layout [3–5].

In this regard, this article considers the issues of automated decision-making of technological equipment layout in a workshop.

2 Setting the Task of Equipment Layout

The problem of equipment layout in technological schemes (TS) can be formulated as follows: to determine such a spatial arrangement of equipment of the TS taking into account all the rules, requirements and constraints with a given structure of technological connections and the size of the production facility at which project costs are minimal.

To formalize the problem, we introduce the following assumptions: 1) it is assumed that the building structure is mounted from standard building elements; 2) the machinery size is approximated by parallelepipeds with sides $a_i, b_i, c_i, i = \overline{1, N}$, N - number of machines to be placed.

Notations: $A_i = (X_i, Y_i, Z_i)$ - spatial location of the i-th apparatus, where X_i, Y_i, Z_i - coordinates of the center of the parallelepiped's base; $A = \{Ai|i = \overline{1, N}\}$ - equipment placement option; $T = \{T_J|j = \overline{1, L}\}$ - pipeline tracing option; $S = (X_C, Y_C, Z_C, H_C)$ - variant of the building structure, where X_C, Y_C, Z_C - length, width and height of the workshop, accordingly, H_c - column grid pitch; $h = (A, T, S)$ - layout option, $h \in \mathrm{H} \subset D$, where H - lot of acceptable layout options, D - lot of all possible layout options $D = D_A x D_T x D_S$; where D_A, D_T, D_S - sets of options of equipment placement, pipeline tracing and building construction, accordingly: $D_A = \{D^{q1}|q1 = \overline{1, |n2|}\}, D_T = \{T^{q2}|q2 = \overline{1, |n2|}\}, D_S = \{S^{q3}|q3 = \overline{1, |n3|}\}$, where $|n1|, |n2|, |n3|$ - capacities of sets D_A, D_T, D_S; m - design solution model that highlights the H set, $m : D \to H$ or $\mathrm{H} = \mathrm{m}(D)$.

Taking into account given notations the problem of equipment layout is formulated as:

$$\text{find } h^* = \arg\min\{I(h)|h \in \mathrm{H}| = m(D)\} \tag{1}$$

As the objective function $I(h)$, we take the above costs, including capital and operating costs for pipelines, means of transportation and construction.

3 Information and Logical Model of Equipment Placement Process Within a TS

To formalize the mathematical model of the design solution of the placement problem, we introduce additional notation: h_x, h_y, h_z - dimensions of the construction quadrant; β_x, β_y - dimensions of the cross-section of columns; Δx, Δy, Δz - maximum permissible distances between the device and the wall; $A^r \subset A$ - subset of devices whose placement is affected by the constraint r, $r = \overline{1, n}$.

$$A^r = \left\{ A^{rk} \middle| A^{rk} = \left\{ A_i^{rk} \middle| i = \overline{1, l^{rk}} \right\}, k = \overline{1, k^r} \right\} \tag{2}$$

where n – number of constraints in the mathematical model; A^{rk} - the kth subset of $A^{rk} \subset A^r \subset A$ devices, the placement of which must be carried out in accordance with the constraint r; k^r - number of subsets A^{rk}; l^{rk} - number of devices of the subset $A_i^{rk} \in A^{rk}$ $i = \overline{1, l^{rk}}$ - device numbers.

Then, taking into account these notations, basic requirements to the building structure and equipment placement can be presented in the form of the following conditions:

Condition 1. The maximum permissible dimensions of the workshop are regulated by building codes and rules. For example, the number of floors in a workshop depends on specific features of the projected production line and varies from 1 to 5. The number of spans along the width of the workshop also depends on the production conditions, but, as a rule, it is less than 6.

$$x_{\min} \leq x_c \leq x_{\max}; \; y_{\min} \leq y_c \leq y_{\max}; \; z_{\min} \leq z_c \leq z_{\max} \tag{3}$$

Condition 2. Since the building structure is mounted from standard structural elements and has the shape of a rectangle on the plan, the dimensions of the workshop should be multiples of the size of the construction module.

$$x_c/n_x = y_c/n_y = z_c/n_z = \Omega, \Omega = \{6, 9\} \tag{4}$$

where n_x, n_y - number of spans along the length and width of the workshop, respectively; n_z - number of floors.

Condition 3. The equipment of the considered productions is placed, as a rule, inside the workshop:

$$\begin{cases} a_i/2 + \Delta x \leq x_i \leq x_c - a_i/2 - \Delta x, \\ b_i/2 + \Delta y \leq y_i \leq y_c - b_i/2 - \Delta y, \\ 0 \leq z_i \leq z_c - c_i - \Delta z, \forall i \in A^3. \end{cases} \tag{5}$$

Condition 4. Devices in the workshop can be placed along or across it:

$$(a_i = a'_{i''} \wedge b_i = b'_{i''}) \vee (a_i = b'_i \wedge b_i = a'_i), \forall i \in A^4. \tag{6}$$

Condition 5. To comply with the gravity of material flows, it is necessary that:

$$z_{f_{1l}} - H_{f_{1l}} \geq z_{f_{2l}} + C_{f_{2l}} - H_{f_{2l}}, \forall l : f_{4l} = 1. \tag{7}$$

The following two conditions are obvious, but, nevertheless, they also need to be taken into account for the layout.

Condition 6. Non-intersections of devices with each other:

$$\left[(z_i = z_j) \wedge \{ ((x_i + \lambda_x \frac{a_i}{2} + \delta_x^i) - (x_j - \lambda_x \frac{a_i}{2} - \delta_x^j))\lambda_x < 0, \vee \right.$$
$$\left. ((y_i + \lambda_y \frac{b_i}{2} + \delta_y^i) - (y_i - \lambda_y \frac{b_i}{2} - \delta_y^i))\lambda_y < 0 \} \vee (z_i \neq z_j) \right], \tag{8}$$

where $i, j \in A^6$; $\lambda_x = sign(x_j - x_i)$; $\lambda_y = sign(y_j - y_i)$; (δ_x^i, δ_y^i) - service area of the i - th device .

Condition 7. Non-intersections of apparatuses with building columns:

$$\left\{ (x_i - a_i/2 \geq \left[\frac{x_i}{h_x} \right] h_x + \beta_x/2) \wedge (x_i + a_i/2 \leq \left[\frac{x_i + h_x}{h_x} \right] h_x - \beta_x/2) \right\} \vee$$
$$\left\{ (y_i - b_i/2 \geq \left[\frac{y_i}{h_y} \right] h_y + \beta_y/2) \wedge (y_i + b_i/2 \leq \left[\frac{y_i + h_y}{h_y} \right] h_y - \beta_y/2) \right\} \quad \forall i \in A^7 \tag{9}$$

Condition 8. The placement of a piece of equipment is limited in height. Large-sized equipment, as a rule, is placed on the lower floor. Reaction equipment is installed on the upper floors:

$$(Z_l/n_z) \in \xi^i, \xi^i = \left\{ \xi_j^i \middle| j = 1, P^i \right\} \forall i \in A^8 \tag{10}$$

where ξ^i - a set of floor numbers on which the i -th unit can be installed.

Condition 9. Some of the devices must be placed under each other, which is caused by the nature of the transportation of substances between these devices:

$$z_i \neq z_j, y_i = y_j, x_i = x_j, \forall i, j \in A^9 \tag{11}$$

Condition 10. It is advisable to combine the same type of equipment into specialized units, which increases the convenience of their maintenance:

$$z_i = z_j, \forall i, j \in A^{10}. \tag{12}$$

Condition 11. It is recommended to place some of the equipment in a row (for example, when loading the same type of devices with one transport unit):

$$z_i = z_j, x_i = x_j, \forall i, j \in A^{11}. \tag{13}$$

Condition 12. For free maneuvering of transport on the floors, zones for movement of transport devices should be allocated:

$$z_i = z_j, \left[\frac{|x_i - x_j|}{h_x}\right] = 0, \left|\left[\frac{y_i}{h_y}\right] - \left[\frac{y_j}{h_y}\right]\right| \geq 2, \forall i, j \in A^{13}. \tag{14}$$

Condition 13. Equipment requiring visual quality control of products should be installed in areas with natural light:

$$(y_i \geq y_j, \vee j \in A\backslash A^{14}) \vee (y_i \leq y_j, \vee j \in A\backslash A^{14}), \forall i \in A^{14}. \tag{15}$$

Condition 14. For the repair and maintenance of devices, as well as for the placement of instrumentation panels, auxiliary rooms, etc., zones free of the equipment should be provided in the workshop:

$$\left\{(z_i = z_m) \wedge (|x_i - \tilde{x}_m| \geq \frac{a_i + a_m}{2} \vee |y_i - \tilde{y}_m| \geq \frac{b_i + \tilde{b}_m}{2})\right\}' \vee \{z_i \neq z_m\}, \forall i \in A^{15}. \tag{16}$$

Here $(\tilde{x}_m, \tilde{y}_m, \tilde{z}_m)$ - coordinates of the center of the m-th room, free of equipment placement.

Condition 15. For technological reasons, as well as for the conditions of transport of substances, the distance between some devices should be limited:

$$|x_i - x_j| + |y_i - y_j| + |z_i - z_j| \leq \delta_{i,j}, \forall i, j \in A^{16}. \tag{17}$$

In addition to these, other conditions are also included in the placement model: no intersection of devices with service areas of other devices, with pipelines and a number of other conditions regulating the procedure and the rules for placing of the equipment.

Thus, the task of equipment placement in the workshop is to determine such coordinates $\{(x_i, y_i, z_i), i = \overline{1, N}\}$ of the equipment to be placed and the size of the workshop (X_c, Y_c, Z_c), at which model constraints (2)–(16) are met and the criterion (1) reaches a minimum.

The use of the mathematical model described above in problem-oriented computer-aided design systems allows (by partially modifying the constraints of the model) to simulate a wide range of tasks for the placement of industrial facilities [6–8].

4   Results and Discussion

The support model for decision-making in industrial production design which is proposed by the authors is illustrated by the example of spirit production.

To implement the technology of creating an industrial digital prototype (in this case exemplified by an alcohol production facility) at the first stage we create a layout of the entire production area, including the adjacent territories with the natural ecosystem

using the Twinmotion software, which provides immersive architectural 3D visualization. The program is intended for specialists in the field of architecture, construction and
landscaping [9–13].

Twinmotion is based on the Unreal Engine platform, the power of which is combined
with an intuitive interface. In this software, terrain, roads, and utility infrastructure
systems are designed. The model created in Twinmotion provides direct synchronization
of geometry and BIM information in one click from your SOLIDWORKS, ARCHICAD,
Revit, SketchUpPro, Rhino (including Grasshopper) or RIKCAD models. (see Fig. 1).

The next step is to create a multimedia virtual tour using the 3DVista Virtual Tour
Suite [11]. The created tours can be viewed online and offline on any device without the
need to install any special software or plug-ins. There is also the possibility of remote
discussion of the project and the model via a browser and the possibility of creating an
e-learning system.

The created virtual tour can be used for a variety of tasks: remote discussion of
the designed equipment 3D model; staff training; emergency situations simulation;
completing educational quests on the enterprise territory.

To implement this function, we will use the 3DVista tool - Remote discussion of the
VR space which is implemented through the built-in tool - Live-Guide Tour (see Fig. 2).

Live-Guided Tours allow you to conduct video conferences in addition to a virtual
tour. Thus, the authors propose a technology for creating a digital prototype of industrial
production, used by specialists as a tool at the design and functioning management
stages.

Fig. 1. Visualization of the general view of the digital model of the projected production.

Fig. 2. A fragment of a panorama inside a facility during a remote discussion of collisions that occurred during the equipment placement and the pipeline system tracing.

5   Conclusion

It should be noted that based on the analysis of the problems associated with greenhouse gas emissions in various sectors of the world economy, the authors propose an approach to managing greenhouse gas emissions on the scale of the regional economy, according to which individual enterprises in the region are motivated either to obtain additional profit or to reduce costs, namely: the enterprises of the first group sell the rights to emit GG to others due to the savings received from environmental investments and the achievement of standards, using the most effective technologies; enterprises of the second group buy GG emission rights without attracting investments in environmental protection equipment, and also minimize their costs.

The results of solving the GG emission quotas redistribution problem (1)–(7) for an individual EE are presented in graphical and tabular form, in particular, the information is provided on the amount of purchased/sold quotas, their cost, as well as on the amount of fines for exceeding the GG emission limit for the entire region.

Thus, the authors proposed a tool to stimulate participants of economic activity, the use of which will significantly bring them and the society closer to a low-carbon economy.

References

1. Lämkull, D., Zdrodowski, M.: The need for faster and more consistent digital human modeling software tools. Adv. Transdiscipl. Eng. **11**, 299–310 (2020)
2. Pavlov, D., Sosnovsky, I., Dimitrov, V., Melentyev, V., Korzun, D.: Case study of using virtual and augmented reality in industrial system monitoring. In: Conference of Open Innovation Association, FRUCT 2020-ril, 9087410, pp. 367–375 (2020)

3. Moshev, E., Romashkin, M., Meshalkin, V.: Development of models and algorithms for intellectual support of life cycle of chemical production equipment studies in systems. Decis. Control **259**, 153–165 (2020)
4. Meshalkin, V.P., Gartman, T.N., Kokhov, T.A., Korelstein, L.B.: Heuristic topological decomposition algorithm for optimal energy-resource-efficient routing of complex process pipeline systems. Dokl. Chem. **482**(2), 246–250 (2018). https://doi.org/10.1134/S0012500818100087
5. Egorov, S., Sharonin, K.: Automated decision making in the problem solving of objects layout for chemical and refining industries using expert software systems. Chem. Pet. Eng. **53**(5–6), 396–401 (2017). https://doi.org/10.1007/s10556-017-0353-3
6. Pozdneev, B., Tolok, A., Ovchinnikov, P., Kupriyanenko, I., Levchenko, A., Sharovatov, V.: Digital transformation of learning processes and the development of competencies in the virtual machine-building enterprise environment. J. Phys. Conf. Ser. **1278**, 012008 (2019)
7. Krol, O., Sokolov, V.: Development of models and research into tooling for machining centers. J. East.-Eur. J. Enterp. Technol. **3**, 12 (2018)
8. Alekseev, V., Lakomov, D., Shishkin, A., Maamari, G.A., Nasraoui, M.: Simulation images of external objects in a virtual simulator for training human-machine systems operators. J. Phys. Conf. Ser. **1278**(1), 012008 (2019)
9. Nemtinov, V., Kalach, A., Egorov, S., Nemtinova, Y.: Information support for decision-making in emergency situations. J. Phys. Conf. Ser. **1902**(1), 012080 (2021)
10. Nemtinov, V., Zazulya, A., Kapustin, V., Nemtinova, Y.: Analysis of decision-making options in complex technical system design. J. Phys. Conf. Ser. **1278**(1), 012018 (2019)
11. Virtual Tours in E-Learning, Training and Quizzing. https://blog.3dvista.com/2020/04/27/virtual-tours-in-e-learning-training-quizzing/. Accessed 27 July 2021
12. Introduction to Twinmotion. https://www.unrealengine.com/en-US/onlinelearning-courses/introduction-to-twinmotion. Accessed 27 Sept 2021
13. Twinmotion: Real-time immersive 3D architectural visualization. https://www.unrealengine.com/en-US/twinmotionCarbon emission quotas have risen by 30% in 2020. https://www.interfax.ru/business/742887. Accessed 27 Sept 2021

Automation of the Process for Generation and Publishing Release Notes for Software Applications

Konstantin Figlovsky$^{(\boxtimes)}$ ⓘ and Igor Nikiforov ⓘ

Peter the Great St. Petersburg Polytechnic University, Saint Petersburg, Russia
`figlovskij.ks@edu.spbstu.ru`

Abstract. The software development lifecycle contains a list of steps to be performed to build software application. It includes requirement preparation, high level design, implementation, testing and preparing the final distribution for the end user and customers. Every step has their own efforts for the team of developers and managers. That efforts are worth reducing by automation approaches to reduce final cost of software product. The paper is devoted to reducing the efforts and time on preparation and delivering software releases to the customer by automation of the process for generation and publishing release notes for software applications. The paper contains research and comparative analysis of existing approaches for release notes generation in various existing tools like: ARENA, Chronicler, Atlassian and other. Based on the research new approach is suggested, and its automation workflow is presented in the paper as well as technical implementation details and integration into continuous delivery tool. The results and metrics that shows benefits of suggested approach usage in the process of mobile application development are listed.

Keywords: Automation · Release notes · Continuous deployment

1 Introduction

One of the steps of the software development lifecycle is building and preparing the target product distribution, that is delivered to the end-user or customer [1].

On this step continuous delivery approaches and tools are used to reduce time and team efforts. Fast product distribution preparation has various benefits for both business and company growth [2]. The first and foremost, it lets software developers team adjust continuously and quickly to consumer change requests and modifications in the software — it enables software developers to gain insights from actual use of delivered product [3]. And the second is that software developers have a greater foundation for recognizing if the product is good enough.

Time to Market (TTM) is one of the most essential product development KPIs or metrics [4]. It is the time from the beginning of the software idea development to its final implementation when the target solution is sold to the client. In other words, Time

A. Gibadullin (Ed.): DITEM 2021, LNNS 432, pp. 78–91, 2022.
https://doi.org/10.1007/978-3-030-97730-6_8

to Market is "the time before entering the market". That is obvious that the shorter TTM then less full software costs are and more revenue of the software vendor and the end.

One of the most essential artifacts in the target software product is the release notes [5]. It describes the changes made to the project's latest release, including new features, enhancements, bug fixes, and deprecated features among other things. When a new version of software needs to be released, release notes are always required, especially for enterprise and mobile applications. Product release notes acts as a direct line of communication with customers. Whether new features or bug fixes are included in the release notes, they are a crucial component of informing end users about the changes. Regularly sharing release notes also aids in the setting of customer expectations. Similarly, product release notes can keep customers and users pleased by letting them know that their feedback is considered in product.

Manual creation of the release notes is a time-consuming and error-prone process that includes interaction of developers, product managers and technical writers. Manual work on release notes preparation increases the time to market indicator and therefore can lead to a decrease in company profits.

Despite the relevance of working with release notes, there are not so many in the public domain (relative to other topics), so the purpose of this article is to introduce a novel approach of reducing the complexity and labor intensity of creating release notes by automating the process of preparing and storing metadata of product releases, as well as generating and issuing metadata in a format suitable for continuous deployment. To achieve the goal of automation, it is necessary to solve the following tasks:

1. Explore existing approaches to automating release notes generation including ARENA, Chronicler, Atlassian and other.
2. Conduct a comparative analysis of existing approaches to identify the gap that needs to be automated.
3. Propose an approach to automate the process of preparing and storing metadata for product release notes.
4. Implement the proposed approach in a software tool.
5. Demonstrate the effectiveness and reduction of labor intensity from the use of the chosen approaches.

2   Materials and Methods

2.1   Manual Release Notes Publication Workflow

Before proceeding to the description of the processes of working with the release notes, to avoid misunderstandings, it is necessary to agree on the terms and abbreviations used in the paper.

- "RN" is for "Release notes"
- "Distribution point" is a place where a software product release can be placed. It can be an app store like "Google Play" and "App Store" or some other site;
- "PM" ("Product Manager") is the person responsible for the availability of the latest version of the product at all distribution points;

- "Doc&Loc" team ("Documentation and localization") is one or more persons responsible for the relevance and correctness of texts and translations of materials of a particular application release;
- "TS" ("Translation service") is a place where text from one source language can be translated into many languages. It can be service like Smartcat [18] or similar.

Let us consider the manual approach and workflow for preparing and publishing release notes prior to new version of software application release. The workflow may depend on the software development model and software type (desktop, enterprise or mobile), but in general the workflow is almost the same and can be represented in following steps illustrated in Fig. 1.

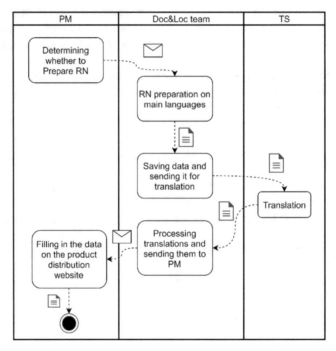

Fig. 1. Manual release notes publication workflow.

1. The product manager decides that it is time to prepare release notes and notifies the Doc&Loc team about the plans;
2. The Doc&Loc team manually prepares RN text for main application localizations and saves it. On that step they closely communicate with the developers and product manager making several interactions on information and knowledge transfer about new features, bug fixes etc. After several iterations the final version of release notes is ready;
3. The Doc&Loc team sends RN to translation service for different languages and waits for the end of it;

4. The Doc&Loc team receives translations, processes, and sends them to PM;
5. The product manager fills RN data in the Distribution Points.

The algorithm described above is not the only way to publish release metadata since everything depends on the specific team and company process. In some cases, the number of participants in the process may be different, or the tasks they perform may differ, but this is not so important within the framework of this work. The most important thing that can be noted is the presence of some manual work that is performed on an ongoing basis and can be automated. Possible automation approaches will be discussed in the following paper sections.

2.2 Existing Solutions and Approaches for Release Note Generation

There are several types of solutions and approaches for automating creation of the release notes. Let's consider them in detail.

The first approach type is a build system's plugins or continuous integration/continuous delivery (CI/CD) systems that automate the publication of applications. As an example of such a system, we can consider the following: codemagic CI/CD [6], Buddy CI/CD [7], Gradle Play Publisher [8] etc. That solutions allow to automate publication of materials for mobile applications. The way these solutions work is simple. The user configures the parameters for CI/CD in the format described for a particular solution. One of the config parameters is the path or link to the location of the release notes files. After starting the application deployment process, the utility copies information on release notes and transfers it to the product distribution website. Those solutions have several disadvantages. First, they work with only "raw" data, which need to be brought into the desired form by hand. The "raw" format can be different for different stores, which brings more complexity, if you need to manage applications for different platforms like IOS, Android, Chrome OS etc. Another disadvantage is that described solutions do not provide any data storage for release metadata and users should independently think about where and how to store the necessary information. Other disadvantage is that there is a tight binding to assembly tools and it hard to change. It is also worth noting that not all solutions are free, and they cost money.

The second type of solutions for automation of release notes generation is text generation systems. Such solutions receive input data, which can be used to determine which changes are included in the scope of the current release. These can be commits, pull requests, tasks in the task tracker, and so on. After receiving the information, the system processes it and generates the text release notes. A detailed research on existing solutions for the text-generating is given in the M. Ali, A. Aftab and W. Buttt article [9].

Researchers examined 6 existing solutions: Laura Moreno's and her team an approach named ARENA [10], Sebastian's Klepper's a semi-automatic method for generating RNs [11], Chaminda Chandrasekara's proposed methodology for generation of automated RNs for Team Foundation Server (TFS) [12] and some other open-source libraries and plugins [13–15]. Also, in addition to the works considered, it is worth noting the Git Change Classifier (GCC) solution for generating and classifying information about changes in the code based on git commits, proposed by Kaur et al. [16].

Table 1. Comparison of output formats for existing approaches.

Algorithm/Tool	Source's format	Output format
Arena	Git commits, issues, documentation, change log	HTML
Semi-automatic approach	Git commits, issues, and change log	Plain text, Markdown, or HTML
Reno	Git commits, change log	Plain text
Atlassian release notes generator	Git commits, change log	HTML
Chronicler	Git commits, change log, and pull requests	GitHub release
Automatic Release Notes Generation	Git repository, change log and issues	Docx
GCC	Git repository	MS-Excel

As shown in Table 1 all these solutions do not consider the features of the release notes for the mobile applications and are more suitable for either open-source libraries or internal use in the company. Another problem is that the solutions do not consider the localization of the product and cannot work with texts in different languages. Data storage also is not provided as in previous solution.

The third solution is services for storing text. It simply stores texts and may have history of changes, but it not automated. As example it can be Google doc or Git version control system [17].

Thus, a new approach is required that does not have listed disadvantages of existing analogues and has all the advantages.

2.3   Automated Release Notes Publication Workflow

The main idea of the proposed approach is to adopt the manual workflow, presented on the Fig. 1, by adding "Metadata Service", that automate gathering, storing, and processing software application metadata.

Application metadata can contain following information about the product: release notes text, graphic materials, software requirements etc. Release notes is main goal of this work.

Automated release notes publication workflow with metadata service is presented on Fig. 2 and it consists of following steps.

1. The product manager decides that it is time to prepare release notes and starts CI/CD pipeline;
2. When the release is formed, an automatic notification is sent to the Doc&Loc team with the information that it is needed to fill in the release metadata. The notification contains a link to service for working with metadata. More about the service implementation is in Sects. 6 and 7;

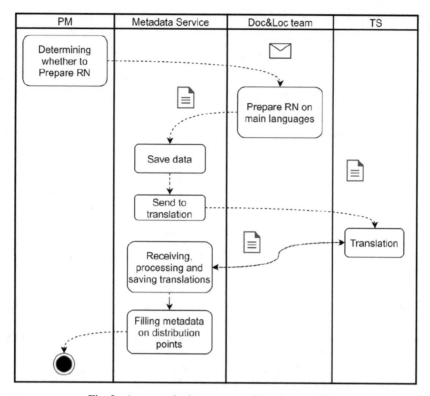

Fig. 2. Automated release notes publication workflow.

3. The Doc&Loc team writes release notes text in the main localizations through the metadata service;
4. The text is automatically sent for translation by metadata service;
5. When the translations are ready, the metadata service automatically saves them;
6. The service automatically uploads metadata on the site with distributed product and notifies product manager.

Based on the above, we can draw an intermediate conclusion that the number of manual actions and interactions between teams has decreased, additional automation has appeared, which allows reducing labor intensity. As mentioned in Sect. 2, the scheme of working with release materials may differ from team to team, respectively, the details of the presented algorithm can be adjusted according to the differences. The most important thing is that the presented approach gives a general understanding of the manual automation process.

2.4 High Level Solution Architecture

To implement the described approach, it is supposed to build a «metadata service». The architecture of the service is shown in Fig. 3.

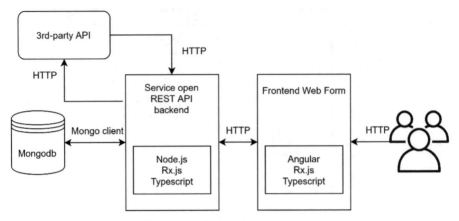

Fig. 3. Metadata service architecture.

The service is divided into 3 main parts:

1. Frontend that contains web user interface and helps user achieve following tasks:

   a. Manage metadata of certain products like supported languages or predefined texts for release notes;
   b. Manage metadata of certain releases like release notes or release task states;
   c. Track rollout status for certain releases;
   d. Track translations status for release notes and send new text to translation service;
   e. Track translations status for release notes and send new text to translation service.

2. Backend, with which Frontend and third-party services interact via the REST API. That module provides following capabilities to the system:

   a. Doing all actions described in frontend capabilities;
   b. Download archives with metadata of certain release.

3. The database that the Backend works with. The level provides capabilities by storing:

   a. Product specific information;
   b. Marketing materials for specific release;
   c. Information about release task states.

The described architecture allows you to automate the collection and storage of release data. Among the functions, it is possible to distinguish the separation of business logic from the user interface, which allows parallel development for several people, as well as the possibility of expanding functionality due to the availability of APIs for third-party services, such as the CI/CD pipeline.

2.5 Technology Implementation Stack

Let us consider the technology stack that is used for implementing the project.

As the main development language Typescript version 3.9.5 is used. Typescript is good for web development and has optional static typing that helps the compiler to show warnings about any potential errors in code, before an app is ever run as opposed to JavaScript. As web application framework Angular version 9.1.4 is used.

Typescript also allows to write high-performance backend logic using Node.js framework version 16.1.0 and Express.js library.

MongoDB version 4.4.3 is used as database. MongoDB has document model that allows virtually any kind of data structure to be modeled and manipulated easily. This property is very valuable for our service because the service needs to work with "documents" whose structure can often change. Also, because MongoDb works with the JSON format, it is convenient for to work with it using the TypeSctipt language.

The code is covered with unit tests to verify the logic and protect solution from introduced errors in the future. As test frameworks Mocha version 8.4.0, Jasmine version 3.5.10 and Karma version 5.0.4 are used. TSLint analyzer version 6.1.3 is used for static code analysis.

The presented technologies are popular and easy to learn, which expands the range of developers who can participate in the development of the service.

2.6 Release Materials Approval Component

One of the most important features of the metadata service is release materials approval feature, that is implemented in its own component.

The component allows following:

– Automate preparation of release notes texts;
– Maintain a list of arbitrary tasks related to a specific release.

The architecture of the component is presented in Fig. 4. Three sub-components are added on the Fronted side:

– The "Check-list" component that implements UI logic for release task managing. The «task» contains information about the actions that need to be performed within a specific release;
– The "Release Notes" component that implements UI logic release notes texts managing;
– The "Approval" component that combine two other components.

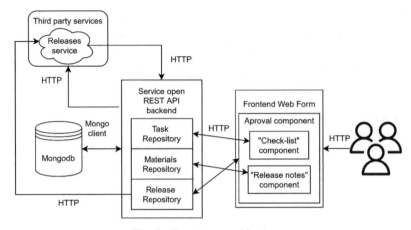

Fig. 4. Component architecture.

Also, there are REST endpoints on the backend side. The endpoints allow to do the following:

– Add, update, delete, and read data about check-list tasks in database;
– Add update, delete, and read data about release notes in database;
– Change and track release state.

The "Check-list" component supports an arbitrary list of tasks for the release. The task can be blocking for release publication and non-blocking. The task can be automatically added to a release based on criteria. The release is published only if all blocking tasks are completed. For example, we cannot publish release until the "Prepare release notes" task completed. The "Release notes" component can:

– Prepare a version of the texts that are proposed to be used for a specific release;
– Automatically fill in the release text based on one of the proposed options;
– Approve Release Notes;
– Verify that Release Notes compliance application store requirements.

Thus, described components allows automating release notes processing, and it remains only to calculate the amount of this automation.

3   Results

To check the effectiveness of the metadata service implementation and suggested automated approach, it is required to measure and compare the time spent on manually publishing release data with the time spent on publishing using the service.

As an example, consider the publication of new versions for two mobile applications, let's call them as application "A" and application "I". "A" is an Android application that supports 36 languages and is published in the Google Play store. "I" is an iOS application that supports 27 languages and is published in the "App Store".

Let's start with application "A". Suppose there are 36 files with ready-made texts for release notes, each file is named according to the localization to which it belongs. It is necessary to compare how long it will take to transfer these translations to the Google Play console and to the metadata service. The results of the five measurements can be seen in Fig. 5.

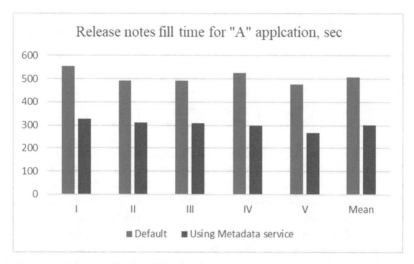

Fig. 5. Measurement results for "A" app.

As a result, it turned out that for Google Play, the average fill time is 508.84 s (14.1 s per locale), and for the metadata service it is 301.94 s (8.38 s per locale). The increase in the speed of work when using the metadata service calculated according to formula 1 was approximately 41%.

$$Growth = \left(\frac{T_{manual} - T_{metadata}}{T_{manual}}\right) * 100\% \tag{1}$$

Now consider the application "I". Suppose there are 27 files with ready-made texts for release notes, each file is named according to the localization to which it belongs. It is necessary to compare how long it will take to transfer these translations to the "App Store" and to the metadata service. The results of the five measurements can be seen in Fig. 6.

As a result, it turned out that for "App Store", the average fill time is 449.304 s (16.6 s per locale), and for the metadata service it is 239.852 s (8.9 s per locale). The increase in the speed of work when using the metadata service calculated according to formula 1 was approximately 47%.

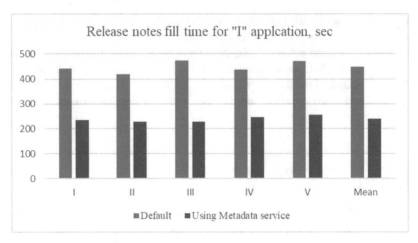

Fig. 6. Measurement results for "I" app.

Most likely, the reason for speeding up the filling of release notes is the differences in the user interfaces of the solutions under consideration (Fig. 7). In the "Google Play" console, the user needs to enter data into the plain-text template with the following format "<language>text</language>", this causes some inconvenience when copying translations from text files. In the "App Store Connect" console, when filling out information about "What's New", you must constantly use the drop-down menu to switch localization, which also causes delays in operation. At the same time, the interface of the "Release Notes" component consists of consecutive text edit boxes for each localization, which avoids the delays that were mentioned earlier.

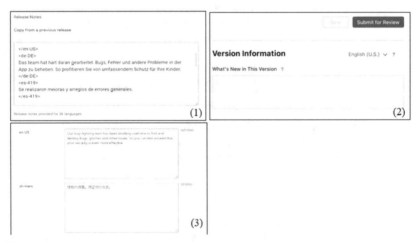

Fig. 7. Release notes inputs in user interfaces for: (1) – "Google Play", (2) – "App Store Connect", (3) – "Metadata service".

Let's try to calculate the amount of work that saves the implementation of the "meta-data service". Consider a random company, which develops N_a number of applications for android, N_i number for iOS. Suppose the management wants to release R_i number of releases for each product per month, while it will be necessary to write release notes for L_i locales. The time for translation of one locale will be considered as T_a for Android applications and T_i for iOS applications Then to calculate the total time spent during the month, you can use the formula 2:

$$Time = \sum_{i=0}^{N_a} (R_i * L_i * T_a) + \sum_{j=0}^{N_i} \left(R_j * L_j * T_i \right) \tag{2}$$

To calculate the amount of work saved, it will be enough using formula (2) to calculate the difference between the time of filling in release notes without using the metadata service and using it. As an example, consider a random company named "K", which develops 20 mobile applications, 10 for Android, 10 for iOS. Suppose the management of "K" wants to release two releases of each product per month, while it will be necessary to write release notes for 30 localizations. Approximately how long it takes to fill out the release notes was measured above in the text. As a result, about 18444 s will be spent without using the service, and about 10200 s with its use. In total, the company "K" will save about 2.3 person-days per month. Of course, for companies with fewer products or supported localizations, the result may not seem impressive enough, but as they say every little helps. In addition, the service can be improved in the future to further reduce the time spent working with release notes.

Also, to determine the effectiveness of the presented solution, it makes sense to count the number of automated steps. The initial solution required 6 manual steps, and there is no automation at all. As a result of introducing metadata service and «approval» component 3 automated steps are added. Manual steps are reduced by 1. A comparison of the steps can be seen in Table 2 below:

4 Conclusion

Thus, in the course of the work, a component for approving release materials for the metadata service has been developed. Component allows to: Prepare and approve materials for release, pre-fill materials from ready-made templates, keep track of an arbitrary list of tasks.

As a result, the number of automated steps was increased by 3 in the entire chain. The number of steps performed manually has been reduced by 1. The reduction in the time for filling out release notes due to the use of the developed component can be more than 40%.

It is also worth noting that the developed service has prospects for further development, which will allow using the introduction of new automated steps to reduce the amount of manual work and further accelerate the release process.

Table 2. Manual and automated steps statistic for approaches.

Approach	Manual steps	Automated steps
Without Metadata service	1. Communication between the manager and the technical writer 2. Storing metadata 3. Sending text for translation 4. Tracking the readiness of transfers 5. Uploading data to the app store	-
With Metadata service	1. Communication between the manager and the technical writer (with small workload) 2. Sending text for translations 3. Tracking the readiness of transfers 4. Entering texts into the service (can be automated with a predefined filling)	1. Storing metadata 2. Pre-filling texts 3. Uploading data to the app store

References

1. Conger, S.: Software development life cycles and methodologies: fixing the old and adopting the new. Int. J. Inf. Technol. Syst. Approach **4**(1), 1–22 (2011)
2. Ambler, S.: Introduction to disciplined agile delivery. s.l.: project management insT (2020)
3. Drobintsev, P.D., Kotlyarov, V.P., Nikiforov, I.V., Letichevsky, A.A.: Incremental approach to the technology of test design for industrial projects. Autom. Control. Comput. Sci. **50**(7), 486–492 (2016). https://doi.org/10.3103/S014641161607004X
4. Time To Market: What it is, why it's important, and five ways to reduce it | TCGen. https://www.tcgen.com/time-to-market. Accessed 13 Oct 2021
5. Abebe, S.L., Ali, N., Hassan, A.E.: An empirical study of software release notes. Empir. Softw. Eng. **21**(3), 1107–1142 (2015). https://doi.org/10.1007/s10664-015-9377-5
6. Nevercode: Codemagic CI/CD. https://docs.codemagic.io/flutter-publishing/publish-release-notes/. Accessed 14 Oct 2021
7. Buddy: Buddy CI/CD. https://buddy.works/. Accessed 13 Oct 2021
8. Saveau Alex: Gradle Play Publisher. https://github.com/Triple-T/gradle-play-publisher. Accessed 15 Oct 2021
9. Ali, M., Tarar, M.I.N., Butt, W.H.: Automatic release notes generation: a systematic literature review. In: Paper presented at the Proceedings - 2020 23rd IEEE International Multi-Topic Conference, INMIC 2020 (2020)
10. Moreno, L., Bavota, G., Di Penta, M., Oliveto, R., Marcus, A., Canfora, G.: ARENA: an approach for the automated generation of release notes. IEEE Trans. Software Eng. **43**(2), 106–127 (2017)
11. Klepper, S., Krusche, S., Bruegge, B.: Semi-automatic generation of audience-specific release notes. In: Paper presented at the Proceedings - International Workshop on Continuous Software Evolution and Delivery, CSED 2016, pp. 19–22 (2016)
12. Chandrasekara, C. Effective release notes with TFS release. In: Beginning Build and Release Management with TFS 2017 and VSTS. Apress, Berkeley (2017)

13. Automated Release Notes for Jira. https://marketplace.atlassian.com/apps/1215431/automated-release-notes-for-jira?hosting=cloud&tab=overview. Accessed 10 Oct 2021
14. Michael Strickland, A.C., Fischer, A., Canaday, A.: Chronicler. https://github.com/NYTimes/Chronicler. Accessed 13 Oct 2021
15. Release-notes-generator. https://pypi.org/project/release-notes-generator. Accessed 14 Sept 2021
16. Kaur, A., Chopra, D.: GCC-Git Change Classifier for Extraction and Classification of Changes in Software Systems. In: Hu, Y.-C., Tiwari, S., Mishra, K.K., Trivedi, M.C. (eds.) Intelligent Communication and Computational Technologies. LNNS, vol. 19, pp. 259–267. Springer, Singapore (2018). https://doi.org/10.1007/978-981-10-5523-2_24
17. Voinov, N., Rodriguez Garzon, K., Nikiforov, I., Drobintsev, P.: Big data processing system for analysis of GitHub events. In: Paper presented at the Proceedings of 2019 22nd International Conference on Soft Computing and Measurements, SCM 2019, pp. 187–190 (2019)
18. Smarcat: All-in-one localization platform. https://www.smartcat.com. Accessed 16 Oct 2021

Process Mining for User Interactions with Russian Wikipedia

Kirill Zabelin$^{(\boxtimes)}$ ⓘ, Angelina Barkanova ⓘ, Daniil Ageev ⓘ, Konstantin Figlovsky ⓘ, and Igor Nikiforov ⓘ

Peter the Great St. Petersburg Polytechnic University, Saint Petersburg, Russia
zabelin.kv@edu.spbstu.ru

Abstract. This article is focused on process mining for knowledge and hidden patterns extractions from Russian Wikipedia based on user interaction data with the resource. The list of analytical tasks for knowledge extraction from Wikipedia is presented and includes: statistics generation, identify anomaly patterns and attempt to establish a link between these anomalies and actual events based on news topics. Initial dataset and approach to handle the dataset is covered in the article. The paper contains a research of existing architecture and technologies that are mainly focused on big data processing and the most suitable architecture is selected to be implemented in the Wikipedia data processing tool. Proposed software solution is described and its usage shows the results of analytical tasks execution. Conclusions on what topics Russian-speaking people are educated in the 2020 year, what events influenced this and other discovered findings are shown at the results with applying of visualization of complex data approaches.

Keywords: Process mining · Big data analytics · Wikipedia · ElasticSearch · Kibana · Spark

1 Introduction

With approaching end of 2020, arguably one of the most intense years in terms of highly discussed worldwide events, it is especially intriguing to look at what people wanted to get knowledge of during those events.

One of the main resources that people reach to for overview on a topic they are interested in is Wikipedia, a free online encyclopedia that is maintained by users all over the world.

Unfortunately, there are not many works and open source software solutions related to Wikipedia process mining and data extraction that may give community realistic insights on what really people were interested in 2020.

The paper is devoted to the development of the software solution, that helps to understand the insights from Wikipedia based on formulated tasks. The distinctive features of that software are: proposed architecture that allows to handle big volume of data in reasonable amount of time with the supported approach of horizontal scaling and easy expansion by adding new analytical tasks.

© The Author(s), under exclusive license to Springer Nature Switzerland AG 2022
A. Gibadullin (Ed.): DITEM 2021, LNNS 432, pp. 92–103, 2022.
https://doi.org/10.1007/978-3-030-97730-6_9

The main point of data mining research is to align popular pages with worldwide events, see how the news impacted number of visits to wiki-pages and understand general interests of Russian-speaking population in 2020.

2 Wikipedia Data Process Mining

2.1 Related Works and Research on Wikipedia Data

During the research of the subject area, two works are considered that also analyze similar dataset for the analogous purposes.

In the work of Zhejiang University [1] a tool called Visitpedia is presented, which allows to detect and analyze social events based on Wikipedia visit history, as well as examine how these events change over time. This paper addresses the issues of detecting various types of anomalies in the data, as well as how they change over time. It also covers the topic of compiling a set of keywords for a specific topic in Wikipedia, for the subsequent search for correlations with real events.

Another article submitted by Batuhan Bardak and Mehmet Tan [2] explores the topic of prediction of influenza outbreaks based on two different sources (Google searches and Wikipedia access logs) using linear regression models. The authors described their method of data preprocessing and algorithm that, using the data, can predict amount of influenza cases in hospitals five days in advance.

In contrast to the existing work, in this paper we place greater emphasis on how to use existing tools to work with the same dataset on various analytical tasks with the least effort, using all the advantages of existing platforms. Furthermore, this work explores how correlations can be found between anomalies and news articles and compares the effect caused by various events across the world.

2.2 Proposed Analytical Tasks

In the work we propose to extract knowledge and data in the following ways:

- Regular statistics;
- Complex approaches to insights extraction.

Into the tasks with regular statistics the following items are included.

1. Compilation of the tag cloud of the most popular articles;
2. Tracking the change in the number of visits to certain articles;
3. Comparison of the number of visits between articles with similar topics (comparison of articles in programming languages).

In addition, for tasks which are beyond regular statistic information, we include tasks similar to finding correlations between anomalies in data and real events we use our custom algorithm.

The main idea of these algorithm for specific Wikipedia topic is:

1. Find anomalies in daily number of page views. For this we find all local extremums and see if number of page views for three days before were quite different from extremum values (by default we check if number of view was lower in 3 times, but it is settable coefficient).
2. Create set of keywords based of Wikipedia page name and content. Every word in name became a keyword as well as the most frequently used words in the page content. Prepositions and some common words are ignored.
3. Get news topic from open sources for time period which is correlate with anomaly time.
4. Find new topics which contain keywords. The new topics which contains these keywords considered as sought.

2.3   Data Source and Format

For solving proposed analytical tasks, it is required to obtain the data for process mining. Wikipedia has several well-documented accessible data stores. For this research, we used Pageviews analytics dataset from dumps.wikimedia.org. All analytics datasets at this site are available under the Creative Commons CC0 dedication, which means that they can be used freely even for commercial purposes.

Data in Pageviews analytics dataset is stored in folders that are separated by year, then by month. Twenty-four.gz archives are generated each day, one for every hour. The records in file that is stored within.gz archive all follow the same format: domain_code page_title count_views total_response_size. First three parameters are the object of our interests, as based on them we can calculate all needed metrics to see trends in page viewing.

3   Software Solution for Data Analysis

3.1   Comprehensive Analysis of Software Architectures for Process Mining

Before describing proposed software solution for implementing analytical tasks, the existing and possible options for implementation should be considered. Table 1 shows a comparative analysis of existing solutions.

Based on the presented Table 1, we can conclude that the ELK-stack provides the ability to efficiently store data, search for it, and also provides embedded tools for analytics, visualization and data processing [5]. Which partially covers our tasks. However, the ELK-stack does not provide the ability to run custom tasks, any custom solutions that use the ELK-stack are separate programs that interact with the ELK-stack through the API [8]. This means that if you need to horizontally scale data processing that occurs outside the ELK-stack, a good solution may be to use an existing platform like Spark or Hadoop, which allows you to run various custom jobs inside these platforms and easily scale them [7, 9].

Since the ELK-stack already uses its own storage (which is part of ElasticSearch), using Hadoop here does not seem to be the best solution, since it stores intermediate results in its own HDFS storage. Spark, in contrast, stores intermediate results in memory.

Table 1. Comparison of platforms for big data tasks.

Criterion	ELK	Hadoop	Spark
Complexity of setting up the environment	Low	High	Middle
Flexibility and extensibility for new analytical tasks	Limited by Kibana features, which is basically suitable for statistic and real-time data	Doesn't provide analytic tools by itself	Doesn't provide analytic tools by itself
Horizontal scaling support	Yes	Yes	Yes
Availability of in-line ETL tools	Yes (Logstash)	Yes	Yes
Built-in storage	Yes (Elasticksearch)	Yes (HDFS)	No (can use HDFS)
Native development language	KQL for Kibana, Any other language which can connect to ELK stack through API	Any language	Java, Scala, Python and R

Which is usually faster, and we also avoid using of two different storages [10]. Therefore, in this work, we decide to use a combination of solutions from the ELK infrastructure and Spark. Since such an architecture allows effectively solve a wider set of analytical tasks, which includes the use of Kibana for statistics tasks and the use of Spark for custom analytics tasks with the possibility of horizontal scaling [4].

To unite two different infrastructures (ELK and Apache Spark) we use ElasticSearch-Hadoop Connector, which makes it easy to use ElasticSerch's storage in Hadoop or Spark jobs, instead of HDFS.

3.2 Proposed Architecture

Full architecture of the developed data analytics project is shown in Fig. 1.

To download the data from the data store, a connector is implemented in Go language. The choice of language to work with large amount of data is crucial, so Go is used because it provides good mechanisms to write concurrent programs, therefore speeding up and optimizing the process of data retrieval. To achieve a good resource utilization (both network and CPU), process, of downloading data and send them to the local ElasticSearch infrastructure, is organized into pipeline. Scheme of connector and its integrations is shown in Fig. 2. Each blue component is a separate goroutine.

As a parameter connector is getting time period for which we want to collect the logs. Before downloading we dynamically create links of a following format: https://dumps.wikimedia.org/other/pageviews/<year>/<year>-<month>, based on the time period and sending GET request to each of the links. Each response contains HTML

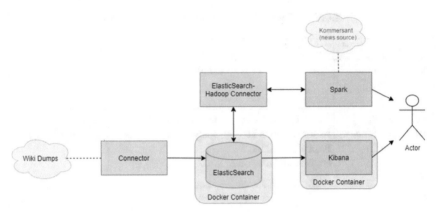

Fig. 1. Data analytics solution architecture

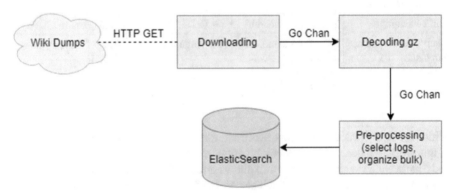

Fig. 2. Data extraction via connector.

which can be parsed to get a list of .gz files which can be downloaded. Another threads, that are not used in data extraction, unpack and process the archives that are now stored in RAM. An important note: the speed of downloading should be the same or less than the speed of processing files, otherwise this can result RAM overload. There are 150.000 records with domain_code = ru on average in one archive and only those strings were extracted from the files. These records are grouped in bulks with total size less than 1MB and are being sent to Elasticsearch (ES) using special ES client for Go (import ("github.com/olivere/elastic/v7")). Each record contains information about date, page heading, number of visits per hour and region. Addition of 'region' field is necessary to analyze other domains in the future if needed.

The retrieved data is processed and saved in Elasticsearch (ES). ES is an open-source search engine that supports distribution and multitenancy [3], which is important for working with big data, as by the nature of this field of knowledge, it is almost impossible to support a full working infrastructure on only one physical device. Data in ES is stored in logical places called indices, a remote analog of tables in relative databases. An index is automatically created during first data transfer from the connector to the ES server.

Each record is now stored at the index as a document (they can be presented in JSON format) with unique identifier. Main advantage of this type of data storage is even if different documents have different number of fields with different types of values, ES easily works with them and provides fast and almost real-time search. There should be said that seemingly the same fields with different type of values are processed as separate fields. This is done to correctly manipulate the data. For the created dataset fields for all documents are the same: date, page title, number of visits and region, however flexibility of developed solution allows for future additions and improvements, as well as expansion of research field.

To parallelize work we use Spark, that is connected to ES through a special open-source ES-Hadoop-Connector. Each Spark job takes the part of data equally.

3.3 Complex Data Visualization

For data visualization tasks (line a regular charts) Kibana is used for this project. Kibana is a visualization platform that integrates with Elasticsearch using REST API. As Kibana and Elasticsearch are both part of Elastic Stack, they are easily compatible with each other, allowing for smart and convenient visualization of big data [6]. In this project, Kibana runs in separate container.

KQL (Kibana Query Language) is a powerful tool that allows using a special query syntax to indicate what data and how they should be visualized on the chart, as well as filter by certain criteria. In this work, KQL was used in conjunction with Kibana's GUI. Through the GUI we determine how the data should be reflected, and with the help of KQL, the filtering of the data was mainly carried out.

4 Process Mining Results

4.1 Tag Cloud for the Year

The first scope that is examined is the most popular pages throughout 2020. In this work, logs are collected from end of December 2019 to beginning of November 2020. This information is shown as tag cloud in Fig. 3, the bigger the word – the more times the page was opened. Service pages of Wikipedia were omitted to give a fuller perspective on users' interests.

The most visited Wikipedia page is 'YouTube' – it is opened more than 3.5 million times during the year. Other online resources were a point of interest of Russian-speaking audience as well, such as RuTracker.org and BitTorrent and related file format.torrent. This can be attributed to the fact that Russia is in the list of countries with most piracy rates.

There are several pages that are related to the pandemic: «Пандемия COVID-19» ('COVID-19 pandemic') and «Распространение COVID-19 в России» ('COVID-19 pandemic in Russia'). Pandemic-related pages will be closer analyzed in section B.

Other type of the most visited pages is geographical, they are dedicated to Russia, Moscow, Saint-Petersburg and, surprisingly, the most popular in the list – Kislovodsk, a spa city in the North Caucasus region of Russia. We attribute such popularity to closure of

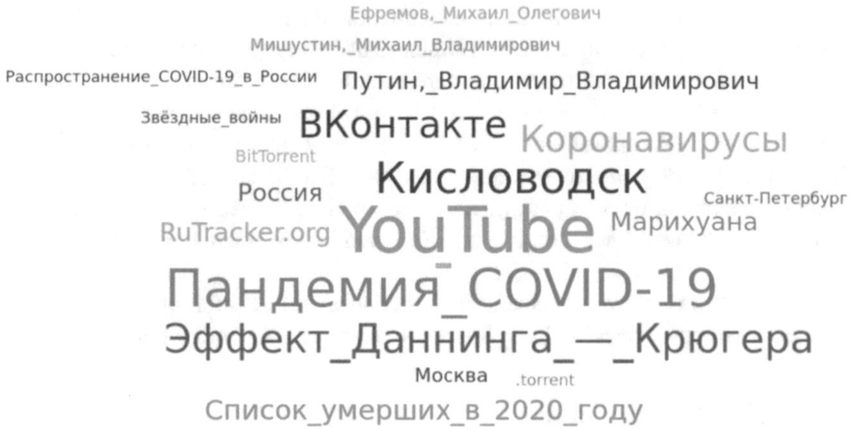

98   K. Zabelin et al.

Ефремов,_Михаил_Олегович

Мишустин,_Михаил_Владимирович

Распространение_COVID-19_в_России   Путин,_Владимир_Владимирович

Звёздные_войны   ВКонтакте   Коронавирусы

BitTorrent

Россия   Кисловодск   Санкт-Петербург

RuTracker.org   YouTube   Марихуана

Пандемия_COVID-19

Эффект_Даннинга_—_Крюгера

Москва   .torrent

Список_умерших_в_2020_году

Fig. 3. The most viewed pages for 2020.

borders due to the pandemic. Many people in Russia travel abroad during their vacations, but without such opportunity a lot of them have decided to explore Russia's resort towns. The 'vacation theory' is also confirmed on the chart in Fig. 4 – there is a sudden growth in July-August, however, there were no major news about the city in this timeline.

The last major group of the most viewed pages is dedicated to famous people: Vladimir Putin, the President of Russia; Mikhail Mishustin, the Prime Minister of Russia who was appointed in January 2020; Mikhail Yefremov, a famous Russian actor who was sentenced to 8 years in prison as the culprit of a car accident that caused death of the driver of the other car. This incident was widely discussed in the Russian media in 2020. All mentioned people were in many headlines, so it is expected to see them in the most viewed list.

One of the most popular results that came out of nowhere is «Эффект Даннинга-Крюгера» ('Dunning–Kruger effect'). This page was visited around 2.5 million times during the year. We cannot correlate such demand with any news or events, maybe Russian-speaking people just became more interested in psychology during lockdown.

4.2   How Fast News Spread: Case Study

In a digital society, it takes a small amount of time to deliver news from one source to another and to final recipient. In this study, we took a couple of major events for Russian-speaking community for which start time can be specified and checked hourly views of pages related to the topic.

The occasion that had been examined was death of a famous and beloved Russian actor Boris Kluyev at the age of 76. The first news about the actor's demise broke around 23:00 MSK on 1st of September 2020. As can be seen from the Fig. 5, in the first hour after the news were published almost 40.000 people have visited the page about the late actor. There can also be a seen a second peak at around 09:00–10:00 MSK. This one could have been caused by people who had gone to bed before news broke, so they got to know them only in the morning.

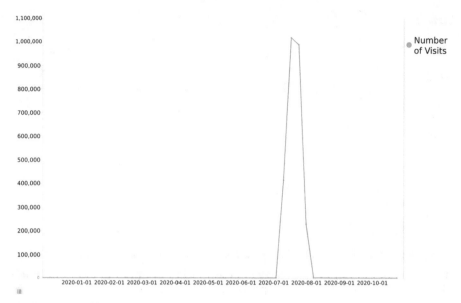

Fig. 4. Visits to «Кисловодск» page. Vertical - the number of views. Horizontal - date per month.

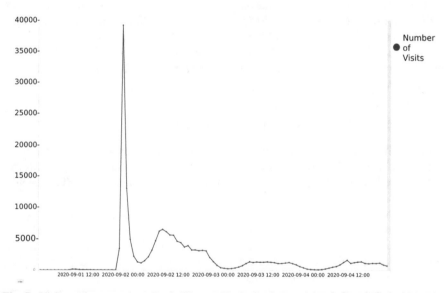

Fig. 5. Visits to the page about Boris Kluyev. Vertical - the number of views. Horizontal - date per week.

4.3 Programming Languages Popularity

With growing interest to IT industry, it is important to research what programming languages are the most popular among people. Most visited pages for programming languages in 2020 are shown in Fig. 6.

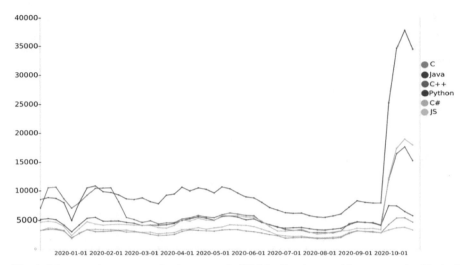

Fig. 6. Most popular programming languages in 2020 for Russian-speaking community. Vertical - the number of views. Horizontal - date per month.

The most searched language turns out to be Python, then C++, then JavaScript. Popularity of Python can be explained by its programmer-friendliness and wide possibilities, as well as usage in modern areas of knowledge such as big data analytics.

There can also be seen a rise for each language starting in October – this possibly correlates with academics; the students could be researching the languages to do their assignments.

4.4 Correlation Between News and Data Anomalies

The search for a connection between real events and sharp spikes in the number of visits of a particular page is currently implemented in the form of a console program that launches Spark jobs, the algorithm of which was described above and a print table with a possibly related new topic.

The table has 4 columns. The first displays the date of the burst. The second displays the ratio between the number of visits at the peak of the spike and that which was before it. The third displays the titles of news articles that were published three days before and three days after the peak of the spike and which potentially can be related to this spike. The fourth column contains links to these articles.

The example is shown in Fig. 7. It displays the result of the algorithm that was run in order to analyze bursts of views and possibly related news related to the Wikipedia page about the Russian president. The table shows that during 2020 there were 6 large bursts of varying strength. Most of them are associated with the publication of certain laws. Also, for some bursts, the algorithm was unable to find related news.

date	popularity rate	new title	new link
2019-12-19	3.070704057279236	Беглов отчитался о планах по ремонту Боткинской больницы после вопроса Путину	https://www.kommersant.ru/doc/4199483
		Путину доложили о стрельбе на Лубянке	https://www.kommersant.ru/doc/4199322
		Мэрия пообещала отремонтировать лифты в доме ветеранов в Сочи после вопроса Путину	https://www.kommersant.ru/doc/4199297
2020-01-15	4.49085123309467	Путин создал рабочую группу по подготовке предложений о поправках к Конституции	https://www.kommersant.ru/doc/4220883
		Путин подписал указ об отставке правительства	https://www.kommersant.ru/doc/4220787
		Путин предложил главу ФНС Мишустина на должность премьер-министра	https://www.kommersant.ru/doc/4220776
		Путин подпишет указ о проведении голосования о поправках к Конституции	https://www.kommersant.ru/doc/4220720
		Медведев станет заместителем Путина в Совбезе	https://www.kommersant.ru/doc/4220712
2020-03-10	5.3348148148148145	Путин на встрече с Медведчуком поддержал идею парламентского измерения «нормандского формата»	https://www.kommersant.ru/doc/4283943
		В Москве проходят пикеты против обнуления сроков Путина	https://www.kommersant.ru/doc/4283922
2020-04-24	27.351500600240097		
2020-10-06	3.281921618204804	Путин одобрил идею Миронова индексировать пенсии работающим пенсионерам	https://www.kommersant.ru/doc/4520801
		Путин готов поставить вакцину от COVID-19 на Украину	https://www.kommersant.ru/doc/4520672
		Путин назвал сентябрь наиболее удобным месяцем для единого дня голосования	https://www.kommersant.ru/doc/4520641
		Путин создал Фонд защиты детей	https://www.kommersant.ru/doc/4520629
2020-10-07	3.826693391115926		

Fig. 7. The example of a correlation table.

4.5 Software Solution Characteristics

We covered the analytics tasks results, but it also worth speaking about the software solution characteristics.

For our testing we use following hardware and software system:

CPU: Intel Core i5-7300HQ
RAM: 16 GB
Network: 100 Mb/s
Amount of nodes: 1

Table 2 shows timing spent on the task execution with notes on complexity for each one.

Based on the table we can see that on the same data set the tasks with different complexity are executed almost the same time. It shows that created software can handle that amount of data in reasonable time.

In case of creation of dataset with more volume there is a capacity to handle the provided data. But if there is not enough hardware capacity, then the approach of horizontal scaling may take place and the software solution should not be rewritten.

In addition, we use following approaches to prove that the software is accurate and has no issues and bugs within the formulated requirements:

– Unit testing, to test individual functions in the connector logic, as well as in the logic of finding correlations between anomalies and real events;
– Functionally testing of connector to make sure the converted data which are stored in ELK stack are not distorted and contain the same information that is contained in the raw source data.

Table 2. Timing spent on the task execution.

Task name	Time (seconds)	Complexity notes
Plotting with Kibana and KQL	~90	Plotting the processed data using Kibana and KQL. Execution time grows linearly depending on the amount of data
Finding correlations between data anomalies and real events	~100	A distributed algorithm that searches for anomalies and tries to find correlations of anomalies with real events. Execution time grows linearly with the amount of data

5   Conclusion

Visualization of big data related to users' interactions with Internet, and Wikipedia in this case, helps to find behavioral patterns of control groups, analyze impact of current or past events on society and learn what the group now has a knowledge of.

The findings of our research have highlighted an important observation – when people learn about something, they tend to not go back to the source of the knowledge if the information is not updated or is no longer interesting to them. This should be taken into consideration for any Wikipedia page - the information there should be clear, concise and correct to give users an idea of some topic.

The same method of the research can be applied to other domains in order to see trends and behavioral patterns of other nations. The software solution that has been created in the work can be easily adopted to other data sources as well.

References

1. Sun, Y., Tao, Y., Yang, G., Lin, H.: Visitpedia Wiki article visit log visualization for event exploration. In: 2013 International Conference on Computer-Aided Design and Computer Graphics, pp. 282–289 (2013)
2. Bardak, B., Tan, M.: Prediction of influenza outbreaks by integrating Wikipedia article access logs and Google flu trend data. In: 2015 IEEE 15th International Conference on Bioinformatics and Bioengineering (BIBE), pp. 1–6 (2015)
3. Kuć, R., Rogozinski, M.: Getting Started with Elasticsearch Cluster in Elasticsearch Server, 3rd edn. Packt Publishing, Birmingham, p. 6 (2016)
4. Mhand, M.A., Boulmakoul, A., Badir, H.: Scalable and distributed architecture based on Apache Spark Streaming and PROM6 for processing RoRo terminals logs. In: Proceedings of the New Challenges in Data Sciences: Acts of the Second Conference of the Moroccan Classification Society (SMC 2019). Association for Computing Machinery, New York, pp. 1–4 (2019)
5. Ermakov, N.V., Molodyakov, S.A.: A caching model for a quick file access system. J. Phys. Conf. Ser. **1864**(1), 012095 (2021)
6. Gupta, Y.: Chapter 1: An Introduction to Kibana. Kibana Essentials. Packt Publishing, Birmingham, vol. 1 (2015)

7. Reguieg, H., Toumani, F., Motahari-Nezhad, H.R., Benatallah, B.: Using mapreduce to scale events correlation discovery for business processes mining. In: Barros, A., Gal, A., Kindler, E. (eds.) BPM 2012. LNCS, vol. 7481, pp. 279–284. Springer, Heidelberg (2012). https://doi.org/10.1007/978-3-642-32885-5_22

8. Kononenko, O., Baysal, O., Holmes, R., Godfrey, M.W.: Mining modern repositories with elasticsearch. In: Proceedings of the 11th Working Conference on Mining Software Repositories (MSR 2014), pp. 328–331. Association for Computing Machinery, New York (2014)

9. Garion, S., Kolodner, H., Adir, A., Aharoni, E., Greenberg, L.: Big data analysis of cloud storage logs using spark. In: Proceedings of the 10th ACM International Systems and Storage Conference (SYSTOR 2017), vol. 1. Association for Computing Machinery, New York, Article 30 (2017)

10. Singh, A., Khamparia, A., Luhach, A.K.: Performance comparison of Apache Hadoop and Apache Spark. In: Proceedings of the Third International Conference on Advanced Informatics for Computing Research (ICAICR 2019), pp. 1–5. Association for Computing Machinery, New York, Article 18 (2019)

Application of BIM Technologies as IT Projects for Digital Transformation in Industry

Oksana Aleksandrovna Krasovskaya[1]([⊠]), Vadim Evgenievich Vyaznikov[1], and Alyona Igorevna Mamaeva[2]

[1] Irkutsk National Research Technical University, 83, Lermontov, Moscow 664074, Russia
chigir-1981@mail.ru

[2] Irkutsk State Agrarian University named after A.A.Yezhevsky, 1, p. Molodezhny, Irkutsk District, Irkutsk Region 664516, Russia

Abstract. Construction of industrial facilities is considered to be a challenging and competitive environment that makes the sector strategically important to the infrastructure and buildings, on which most other sectors depend. It is considered to be a major driver of the global economic growth. Development in this sector is the most important prerequisite for successful decisions covering all areas of socio-economic development and restoration of the economic potential of Russia. The construction industry has tremendous opportunities to increase productivity and efficiency through the digitalization, innovative technologies and new construction methods. The accelerated emergence of augmented reality, building information modeling (BIM), stand-alone equipment and advanced building materials are analyzed. With the use of these innovations, construction companies increase productivity, optimize project management procedures, and improve the quality and safety; both in developed and developing countries, the importance of the construction industry is increasing due to the processes of globalization, technological evolution, sector growth. Due to competitive changes, specialists are striving to adopt and implement appropriate management strategies. By using technologies, construction companies can respond to growing demand. As more construction companies integrate technology into their processes, consumers expect greater speed and efficiency. Experts predict that this cycle will continue, the terms will be reduced and customer expectations will rise. Construction management involves the overall planning, coordination and control of the construction process. The goals of project management are to create a project that meets the client's budget and schedule, has an acceptable risk level and is efficient and safe.

Keywords: Industrial production · Construction facilities · Innovative technologies · Construction industry · Building modeling · Building information modeling

1 Introduction

Manufacturing is part of the economy that deals with the design, construction, maintenance, disposal, modulation, modification and destruction of facilities. It is a large

© The Author(s), under exclusive license to Springer Nature Switzerland AG 2022
A. Gibadullin (Ed.): DITEM 2021, LNNS 432, pp. 104–116, 2022.
https://doi.org/10.1007/978-3-030-97730-6_10

sector and a major contributor to environmental changes, both in terms of the design of the built environment and anthropogenic impacts. Thus, it affects economic, social and environmental problems around the world. The construction industry accounts for approximately 10% of GDP and contributes 30–40% of global energy consumption, as well as 20–30% of greenhouse gas emissions (UNEP 2007). Thus, sustainability is becoming increasingly important in the construction of industrial facilities.

For the successful management of this industry, it is extremely important to implement effective management systems that can process various information and project documents.

An effective management system is one of the key factors. Industrial construction projects usually have a portfolio of independent projects. Thus, when designing a management system for construction companies, it is advisable to take into account the project orientation of this business whose main components are the system of business processes and organizational structure.

The investment construction process involves the interaction of a large number of participants: an investor, a developer, a customer, a contractor, a project company, an engineering and survey company, suppliers and government authorities, local authorities, state supervision and examination bodies. Each region of the Russian Federation has its own authorities and requirements.

The construction sector has always been one of the drivers of the Russian economy. The Russian construction industry is extremely conservative and often criticized for the lack of innovations. There is no investment in the construction of industrial facilities, the implementation of innovations, the development of new ideas. Progress has been made only in high-tech sectors such as information and communications technology, biotechnology and nanotechnology. Currently, despite serious efforts to support innovations, the Russian government has not yet succeeded in creating a positive dynamics of the innovation diffusion cycle in the construction industry. Public funds are inefficiently spent, and the construction sector continues to face barriers.

Innovative building technologies can improve safety, efficiency and productivity of large-scale construction projects.

Innovations in the industrial construction sector have spurred impressive advances in the types of buildings. For example, the development of cofferdams and caissons has contributed to the construction of majestic underwater structures. Meanwhile, advances in the tower crane technology have propelled construction upward, opening up opportunities for massive skyscrapers.

After a long construction boom, 2000 was hard for the construction industry that sought to protect workers. The industry always responds to hard times with a heightened focus on innovations; therefore, next year is likely to see further advances in technologies that are transforming the construction industry.

New technologies increase labor productivity and company efficiency. Innovations are risky which usually discourages construction companies from implementing them. However, only those companies that are dynamic and capable of responding to the challenges of globalization will survive. As a result of the speed at which the global marketplace is moving, only those companies that are capable of innovating and managing the associated risks will be leaders. Due to the competition, innovations are one of the opportunities for the development of construction companies.

Constantly intensifying competition and challenges for construction companies combined with the rapid development of information and communication technologies require the digitalization of business processes in the construction industry. Digitalization as a progressive process of widespread application of information and communication technologies is the focus of many construction companies. Improving business models, efficiency, innovation and quality of management decisions, reducing costs, inclusion of customers in construction processes are some benefits that companies can receive due to the digitalization.

The digitalization in the construction sector is a slow but steady process. The digitization is expected to affect all actors in the construction industry - suppliers of equipment and materials, builders, engineering companies, suppliers or promoters. Digital technologies are used in the production, financing, construction, real estate and insurance sectors, education, government, etc. There is no doubt that the forthcoming digitalization will be significant. In this regard, the study of best digitalization practices in the construction industry and recommendations for successful implementation of digital technologies play a crucial role. In the last decade, attention to digital technologies applied in the construction industry has grown rapidly, and major changes are expected. Various aspects of digitization and digital transformation in the construction industry are discussed in a large number of scientific publications. In order for the industry to make full use of the potential of digitization, it is necessary to study and implement best practices for successful digitization in the construction sector.

2 Materials and Methods

New technologies are used not only for survival. They are a driving force of the economy. Given the competitiveness of construction companies and the economy as a whole, investing in innovation is a key factor. The participants in the implementation of new technologies are contractors, investors, companies producing construction products, engineers, architects, technical support providers and contractors, as well as other persons performing trade and installation works.

This paper describes the types of building information modeling (BIM).

3 Results

Currently, the process of implementing new technologies in the construction industry increases labor productivity and efficiency of the company. Innovations are risky, which usually does not encourage construction companies to implement them. However, only those companies that are dynamic, open and able to respond to the challenges of globalization will survive. As a result of the speed with which the global market is changing, only those companies will be successful those are able to implement innovations and manage innovation risks. Due to the market competition, innovation is considered to be one of the opportunities to achieve further growth of construction companies. New technologies are required not only for the survival of construction companies and the entire construction industry. They are a driving force of the entire economy. Given the competitiveness of construction companies and the economy as a whole, it is necessary to invest in innovations. Participants and managers implementing new technologies are contractors, investors, construction companies, engineers, architects, technical support providers and construction contractors, as well as other persons performing trade and installation works.

After a long construction boom, 2000 was hard for the construction industry that sought to protect its workers. The industry always responds to hard times focusing on innovations. Therefore, the next year is likely to see further advances in technologies that are transforming the construction industry.

There is a newer type of innovative technology in the construction industry. Building Information Modeling (BIM) is the process of creating a digital representation of a structure («model») before building it. An accurate representation of the building allows everyone involved in the construction to anticipate difficulties, eliminate risks, determine logistics and improve efficiency (see Fig. 1).

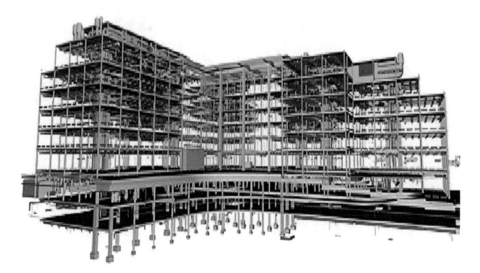

Fig. 1. Building modeling (BIM) in the construction industry.

Replacing the need to create a blueprint before the construction process, Building Information Modeling is one of the biggest advances in the construction industry. This allows you to create a detailed and interactive 3D model that brings together all the data you need about the project. This modeling process eliminates the problem of contractors and clients working in isolation and allows all parties to collaborate effectively.

BIM will gradually become a 5D modeling having features such as time management, cost and quantity calculation.

Building Information Modeling is useful at all stages of construction: before the construction process, BIM reduces the need for future change orders by anticipating problems.

During the construction process, BIM improves communication and efficiency by offering a central hub for up-to-date and accurate reference documentation.

Once built, BIM provides an ability to manage a building throughout the entire lifecycle, providing owners with valuable insight into every detail.

Building Information Modeling may be one of the most important advances in the construction industry as it influences and improves every aspect of the construction process (see Fig. 2).

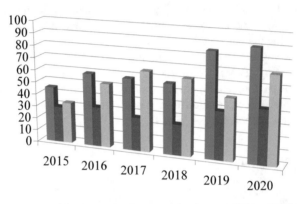

	2015	2016	2017	2018	2019	2020
Investment	45	59	58	57	85	90
Scope of work	29	32	27	25	39	44
Commissioning of housing	33	52	65	62	50	71

■ Investment   ■ Scope of work   ■ Commissioning of housing

Fig. 2. The result of BIM technology implementation in the Russian construction industry (%).

The graph shows that with the BIM used in the construction industry, the share of housing commissioning has increased.

Building Information Modeling (BIM) is the process of creating and managing data about buildings throughout their life cycle, typically using 3D software to improve performance in the design and construction of buildings. It creates a building information model (BIM), which includes building geometry, spatial relationships, geographic information, and properties of building components. The BIM can save time and reduce project costs, as well as increase the overall construction productivity and eliminate design and construction errors.

The digital nature of BIM technologies allows projects to be developed by automated manufacturing software systems used by some component manufacturers (e.g. structural steel, precast concrete, ducting), thereby increasing the potential for more efficient off-site manufacturing. The Design for Manufacturing and Assembly (DFMA) provides production benefits for construction projects, improves safety and productivity and reduces waste.

The modeling of new properties aims to support the design and construction process, which is efficient, safe and consistent with the principles of sustainable development. One of the results of this process is a building information model (BIM).

When applying the technology in the construction industry, several methods can be used to provide support, which can strengthen the control. The BIM technology can reflect the technical quality management process. The BIM technology can provide real-time monitoring of the construction process. The technology is helpful in ensuring the quality of construction.

The factors affecting the quality of projects are people, equipment, materials, methods, and conditions. These factors can be controlled and the quality of construction can be improved. The BIM plays an important role in controlling these factors.

In the construction industry, the head of the construction company and the on-site operator are the main controlling parties. The BIM can help them do their jobs better, more efficiently, and solve construction quality problems. The construction site simulation is an important function of the technology. It helps predict possible quality problems. The operator has a direct impact on the quality of the construction project. The BIM technology can divide jobs at the construction site so that workers play their roles in the construction process, and make their own analysis of key and complex works in order to avoid possible risks.

The BIM technology can record specific information and classify it by different types of materials. For construction projects, the BIM technology can be used in electronic modeling of construction methods; it can be combined with various strengths and weaknesses in the analysis of modeling, it is more consistent with the construction project process.

The BIM technology can be used to develop a program for the demonstration of argumentation, thereby ensuring the quality of construction projects. The BIM technology can determine favorable conditions for the project.

The Building Information Modeling in the construction industry is the process in which information is generated, processed, managed and communicated to project participants. The BIM helps organizations reduces human errors, and prevent project conflicts.

The Building Information Modeling is mainly used in the construction industry. 50% of the construction companies use the BIM. However, more and more companies are seeking to incorporate the BIM into their systems. Many universities have implemented the BIM technology into curricula (see Fig. 3).

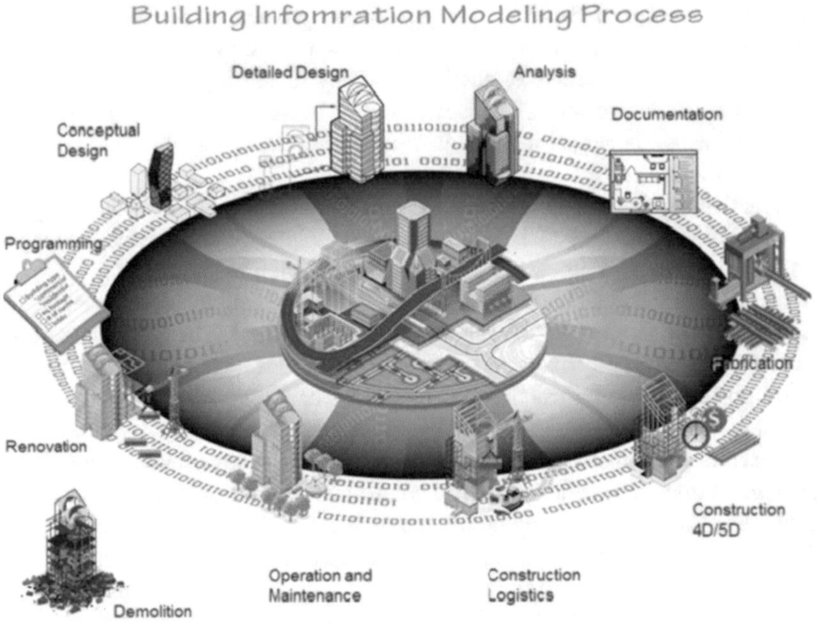

Fig. 3. Building Information Modeling in the Construction Industry.

4   Discussion

BIM stands for Building Information Modeling, which is an information modeling technology.

This technology allows you to model any construction object, including buildings, railways, bridges, tunnels, ports, etc. The similarity between BIM and 3D modeling is that in both cases the construction project is implemented in three-dimensional space. But unlike the 3D model, BIM is directly linked to the database. This model contains data on load-bearing lines, texture of the materials and technological and economic data on buildings. For example, BIM takes into account physical characteristics of the object, options for its placement in space, costs of each brick, ceiling, pipe.

The BIM technology is used to design and document projects in buildings and infrastructure. All building blocks are modeled in BIM. The model can be used to explore possibilities of the project, to create visualizations that will help participants better understand what the building will look like in real circumstances. The model is also used to develop design documents.

BIM allows you to represent a building as a single object in which all the elements are connected and depend on each other. If the system value changes, the remaining data is recalculated. Only the initial data is used to predict future properties and characteristics of the object. In addition, you can use the BIM technology to calculate processes that occur in the object. All information about construction, materials, climate and other factors is digital and the system calculates possible scenarios.

The BIM technologies are at the intersection of different disciplines. With this modeling method, complex data on architecture, design, engineering and economic solutions can be combined into a project, which helps to avoid mistakes, increase the return on the project and its efficiency. Data must be provided in accordance with established standards, be accurate and regularly updated. One of the main advantages of the model is reduced time and costs on the part of the customer and the ability to improve the project at the early stages. The information modeling technology makes the customer a full participant in the construction process. They are able to imagine what the object will be like and make adjustments. The 2D drawing does not give a realistic picture of the future building as much as possible with the BIM model. Sometimes the idea of an architect, designer or customer is not always feasible; in the BIM model, it can be seen at the initial stage. With this modelling, you can see any flaws and possible problems.

It is necessary to create a single information environment that provides immediate access to all data of project participants. The digital BIM model contains work schedules, geographic location, and financial reports. Modern mobile applications can reproduce virtual reality, which recreates the construction object in real conditions. It helps to assess the process of construction, being anywhere in the world.

The information modeling has radically changed interactions of architects, engineers and other construction professionals. The BIM and cloud technologies provide all participants with complete information about the project - materials, technologies, costs and design, logistics, maintenance of facilities during and after construction.

Only rich countries have been using the BIM in the construction industry.

The Building Information Modeling in the construction industry is the process in which information is generated, processed, managed and communicated to project participants. The BIM helps organization reduces human errors, and prevent project conflicts.

The United Kingdom is a leader in the use of BIM. This has become possible due to the government support: since 2016, all budget construction projects have been implementing using the BIM technology. The technology was used by the Justice Department to expand the Cookham Wood Prison in Kent. This allowed the country to reduce capital costs.

In the United States, the Office of Public Services has been using the BIM technology since 2003 for all public building maintenance projects. About 72% of construction companies use BIM to reduce costs. Many US states, universities and private organizations apply BIM standards. Thus, Wisconsin has made the BIM technology mandatory for government projects if their total budget is more than $ 5 million. There are already half a million houses in France designed with the BIM technology. Since 2017, the government has been using the BIM technology for 500,000 homes in the housing sector. The Le Plan Transition Numérique dans le Bâtiment working group is responsible for the BIM strategy to ensure environmental safety and reduce costs.

In Germany, the government also supports the BIM technology. The focus is on commercial and residential buildings with a view to implementing the BIM technology in all infrastructure projects by 2020 (see Fig. 4).

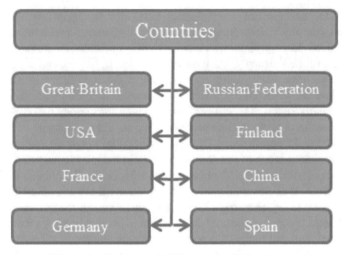

Fig. 4. Applications of BIM technologies by country.

In Spain, the BIM technology has been used for public sector projects from 2018, and since 2019 it has been used in infrastructure projects. A special committee supports the implementation of BIM in the Spanish construction industry.

The Scandinavian countries were among the first to use the BIM technology. For example, Finland started using it in 2002. The BIM technology was used to build complex infrastructures, such as the Helsinki metro line.

Chinese experts from the Atomic Energy Committee have integrated a high-level BIM implementation policy to digitize and disseminate the technology. The BIM technology has become a key element and is used in most projects. GOC has not yet implemented the BIM technology in the construction industry but encourages its application.

In the Russian Federation, the BIM technology appeared later, but the BIM market is rapidly developing. According to PWC estimates (since April 2020), only 5–7% of Russian companies use BIM technologies. In addition, 80% of companies use the BIM technology in the design process and 15% of companies use it in the construction process. Russia has reached a mature level of application of 3D in civil engineering. The next step is to implement 4D in the whole construction process (Synchro Pro and other similar software).

According to analysts, the BIM technology can reduce the number of errors in the project documentation by 40%, planning time - by 20–50%, project verification time - by 6 times, schedule approval time - by 90%, project implementation time – by 50%, time construction time - by 10%, construction and operating costs - by 30% (see Fig. 5).

The BIM market will develop in the coming years. PWC expects it to grow 1.5 times by 2023. The growth is likely to be even more impressive given the new Prime Minister's decree on the use of BIM technology.

In general, even in economically developed countries, BIM is an experimental technology and its implementation is slow for many reasons. However, the digitization is still accelerating, and developers are showing great interest in modern long-term solutions, such as information modeling of buildings.

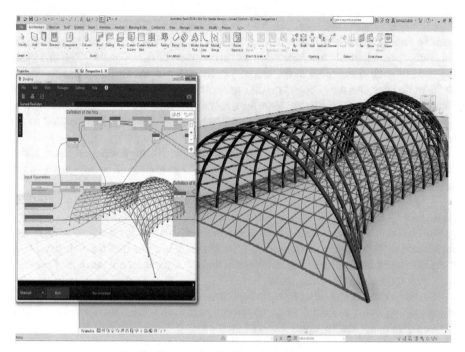

Fig. 5. Construction facility simulation software.

The global trend is transition from three-dimensional modeling (3D) to more complex formats: 4D (3D + construction schedule), 5D (4D + cost), 6D (5D + construction time), 7D (6D + repair/reconstruction periods). These formats are used in the development and implementation of complex and large projects.

The Swiss group Strabag has implemented a series of design projects using the BIM 5D technology (e.g., the seven-storey Siemens office building in Switzerland and the ThyssenKrupp high test tower in Rottweil, Germany). The tower is intended for testing high-speed elevators. The tower has the highest viewing platform in Germany with a panoramic view of the city and suburbs of Rottweil from a height of 232 m. The use of BIM 5D increased the speed of construction of the tower to 3.6 m per day, which allowed for continuous logistics of construction materials. Technologies will significantly reduce the construction time of skyscrapers around the world.

In North America, Europe and developed Asian countries, the BIM 5D format was able to compete with 3D for distribution a few years ago. In Russia, the most promising format is BIM 4D, which can reduce delays in the construction process. For example, it is common that initial calculations of construction materials do not meet real needs. For example, 100-m pipes were purchased, but additional four taps were required. The need for further purchase stopped the construction process. Four-dimensional BIM technology can prevent these situations. You do not have to spend time on changes, so the time and cost of construction can be reduced.

In practice, it is still quite difficult to create a 5D model based on construction costs. The prices in the Russian market are determined by the calculation norms, and the calculation norms are separated from real elements of the model. For example, to install a door, you can buy a box, canvas, walkways, and accessories. In the calculation database, these elements are referred to as "installation of door panels". You can place the costs of each component in the model, but the costs of the group cannot be related to the model because they depend on the elements. It is almost impossible to remove all elements from the cost elements; therefore, it is difficult to use 5D models. 6D models allow the maintenance of buildings and structures, and 7D models can be used in the repair and reconstruction processes.

By the end of 2022, all major projects in Russia will be developed using BIM-technologies; the bill on the mandatory use of BIM-models will be supported in the implementation of government orders, and BIM 4D will be in good position.

According to the UN, by 2050 the world's population will reach 10 billion people. The Global Architecture and Construction Industry (AEC) is responsible for the organization of social and economic spaces of the population, as well as for the preservation and restoration of existing buildings and infrastructure. It is clear that the industry needs more efficient planning and construction methods that meet global needs and create a more rational and stable living space.

The information modeling of buildings increases the efficiency of design and construction, and helps to save data on the works performed. Therefore, interest in BIM technologies continues to grow.

5 Conclusion

New technologies have been developed to manage construction processes. The advantages of traditional civil construction building technologies have been weakened, and new building technologies are being implemented to create a good environment for the development of building technologies.

For the construction industry with increasing construction qualities and scale of construction, innovations influence traditional architectural views. Improved residential construction technologies are used to abandon the shortcomings of the traditional technologies, to preserve efficient technologies and implement innovations, to combine new technologies and old elements with the help of current resources and technical conditions, to adapt to new requirements in the field of modern engineering technologies.

To improve the quality of construction, more effective designing methods are being developed. BIM technologies are one of the most striking innovations in the construction industry. Currently, the main purpose of BIM technologies is to support a construction project using 3D models throughout its entire life cycle. The life cycle of a construction project information model consists of several stages. One of the stages is designing. At this stage, a 3D model is created, an engineering analysis of the construction decision is conducted, a graphical presentation and a tabular presentation of the project are provided.

Thus, building information modeling (BIM) is a special information system that has a numerical description and can be used at the design and construction stages and even during operation and demolition of the facility.

The rapid development of new technologies makes the construction industry change production processes, develop new activities in the management of construction projects. New models of industrial facilities can be put into operation faster.

BIM is a process that simplifies the designing of buildings by using a well-organized system of computer models instead of a separate set of drawings. BIM simplifies the construction process and saves time. Outsourcing services related to the implementation of BIM technologies are becoming increasingly popular.

The development of BIM technologies is a process required to improve the quality of projects developed at the planning, construction and maintenance stages. However, these technologies are difficult to implement at all levels; for small and medium enterprises, the implementation of BIM technologies can be quite expensive. Typical projects can be implemented in 2D, but for complex, large-scale projects it is more expedient to use this information modeling technology.

Another important aspect is the training of qualified personnel through advanced training and by universities. Students need to learn modelling tools and understand all modelling stages. Only in this case, they can become valuable and competitive experts in the labor market. Understanding design processes reduces working hours, avoids unnecessary operations, improves quality and provides a clear view of the project.

References

1. Ai, W., Chi, B.: The impact of capital structure on financial performance of real estate enterprises under deleveraging. In: ICCREM 2019: Innovative Construction Project Management and Construction Industrialization - Proceedings of the International Conference on Construction and Real Estate Management, pp. 759–765 (2019)
2. Alekseeva, T.: Acceleration of the cycle of extended reproduction of the active part of fixed assets in construction. In: E3S Web of Conferences, vol. 33, pp. 03057 (2018)
3. Antipina, O.V., Velm, M.V.: Characteristics of project management in the construction industry of the Russian Federation in modern economic conditions. IOP Conf. Ser.: Earth Environ. Sci. **751**(1), 012072 (2021)
4. Barykina, Y.N., Chernykh, A.G.: The leasing development tools in the construction industry of the Russian Federation. IOP Conf. Ser.: Earth Environ. Sci. **751**(1), 012133 (2021)
5. Cheng, Y., Clark, S.P., Womack, K.S.: A real options model of real estate development with entitlement risk. Real Estate Econ. **49**(1), 106–151 (2021)
6. Coulson, N.E., Dong, Z., Sing, T.F.: Estimating supply functions for residential real estate attributes. Real Estate Econ. **49**(2), 397–432 (2021)
7. Giannetti, A.: Home sales pair counts: the organic metric for trading volume in housing markets. Real Estate Econ. **49**(2), 610–634 (2021)
8. Ilina, E., Tyapkina, M.: Enterprise investment attractiveness evaluation method on the base of qualimetry. J. Appl. Econ. Sci. **11**(2), 302–303 (2016)
9. Krasovskaya, O.A., Chigir, A.E.: Types and methods of application of information technologies in the transport industry of Siberia. IOP Conf. Ser.: Mater. Sci. Eng. **760**(1), 01203 (2020)
10. Krasovskaya, O.A., Vyaznikov, V.E.: The lending efficiency in the construction industry. IOP Conf. Ser.: Earth Environ. Sci. **751**(1), 012152 (2021)
11. Kuznetsov, S.M., Gavrilova, Z.L., Teplouhov, O.J., Avetisyan, B.R., Markova, S.V.: Improving the methods of monitoring and automation and mathematical modelling of railway protection. J. Phys.: Conf. Ser. **1333**(4), 042021 (2019)
12. Liu, C.H., Liu, P., Zhang, Z.: Real assets, liquidation value and choice of financing. Real Estate Econ. **47**(2), 478–508 (2019)

13. Nechaev, A.: Prokopyeva, A: Identification and management of the enterprises innovative activity risks. Econ. Ann.-XXI **5–6**, 72–77 (2014)
14. Nechaev, A., Romanova, T., Tyapkina, M.: Author's toolkit of the state regulation of the development of leasing. MATEC Web Conf. **212**, 09010 (2018)
15. Nechaev, A.S., Zakharov, S.V., Barykina, Y.N., Vel'm, M.V., Kuznetsova, O.N.: Forming methodologies to improving the efficiency of innovative companies based on leasing tools. J. Sustain. Finan. Invest. 1–18 (2020). https://doi.org/10.1080/20430795.2020.1784681
16. Porsev, E.G., Gavrilova, Zh.L.: Modern approaches to water drying in the underground transport system. IOP Conf. Ser.: Mater. Sci. Eng. **560**(1), 012196 (2019)
17. Zakharov, S., Shaukalova, A.: Methodological aspects of optimization of small enterprises in modern conditions of the Russian economy. IOP Conf. Ser.: Mater. Sci. Eng. **667**(1), 012108 (2019)

Managerial Decision-Making in the Sphere of Tourism Under the Conditions of Risk and Uncertainty

Maria Sahakyan[1]([✉]) [iD] and Elena Antamoshkina[2] [iD]

[1] Armenian State University of Economics, 128, Nalbandyan Street, 00025 Yerevan, Armenia
sahakyan_maria@yahoo.com
[2] Volgograd State Agrarian University, 26, Universitetskiy Avenue, 400002 Volgograd, Russian Federation

Abstract. The present paper addresses the justification of the technique of strategic managerial decision-making in tourism industry, which is currently characterized by a high level of risk and uncertainty amidst the COVID-19 pandemic. The authors of the paper would like to identify effective strategies for the development of tourism industry on the example of Russia and the Republic of Armenia. The paper considers four possible strategies for tourism recovery based on expert assessments of the probability of restoration of the pre-pandemic situation in the field of inbound and domestic tourism over the medium term. The strategies included developing tourism clusters and destination management organizations, starting the Ministry of Tourism, preserving the current policy of tourism development. The research uses probabilistic and mathematical decision-making methods based on extrapolation, and provides a solution matrix of possible strategies to stimulate tourism in the post-pandemic period in Russia and Armenia. According to experts, the tourism industry is most likely to recover by 60% over the next 5 years compared to 2019. Wald, Savage, Hurwitz and Laplace criteria were used to evaluate the strategies for a long-term strategy development. The findings suggest that the cluster strategy in the tourism sector will be most effective for Armenia. In the tourism industry of Russia, it is recommended developing the current regulatory system and policy; however, the clustering strategy is also effective, as its implementation risk is low.

Keywords: Tourism industry · Development strategies · External environment · Risk and uncertainty factors · Decision-making models

1 Introduction

Human capital is one of the most important resources of any organization. Managerial personnel play a decisive role in the process of the development and success of any company. Theoretical studies which clarify the role and the content of managerial work have become popular relatively recently. H. Mintzberg studies stand at the origins of these works [1]. In his later research, Mintzberg abandons the definition of "managerial work"

A. Gibadullin (Ed.): DITEM 2021, LNNS 432, pp. 117–129, 2022.
https://doi.org/10.1007/978-3-030-97730-6_11

and uses the term "management" in the same context [2]. Forty years later, Mintzberg tried to dispel the myths and misinterpretations about the activities of managers at various levels, considering management not as a profession or art, like most theorists, but as practice and everyday work, conditioned by the specific capabilities and abilities of each individual manager [3]. Thus, according to Mintzberg's interpretations, management, including managerial decision-making, despite advanced technologies, comes down to personal qualities and experience. At the same time, the latter concepts are very extensive and comprehensive, since they include intuition, knowledge, skills, personal traits and the character of a manager.

Back in the early 1990s, management theorists noticed dramatic changes in the management process; especially related to the process of managerial decision-making and the hierarchical structure of an organization. It was during this period of time that many authors began to note conspicuous trends in the decentralization of managerial decision-making [4]. Simultaneously, they observed a significant weakening of vertical organizational structures [5]. Later, theorists began to pay more attention to leadership issues than to management problems, and the mechanisms for managerial decision-making were mostly dependent on specific leadership styles [6, 7].

Managerial decisions can be made both under the conditions of certainty, and those of risk and uncertainty [8]. To clarify the conditions inherent in these situations, it is important to state that the situation of certainty refers to the short term, when the external environment does not change so rapidly and it is possible to proceed from the assumption that there are no unforeseen circumstances. Under a state of risk, the decision maker can determine the probability of the emergence of alternatives, but under the conditions of uncertainty, the decision maker has information about available alternatives only but can't predict the probability of their emergence.

Since tourism is an economic sector that is very sensitive to changes in the external environment, the situation of certainty cannot be considered as a condition for making decisions in this area. Therefore, the present paper addresses the decision-making toolkit under risk and uncertainty.

Overall, the paper analyzes the methodology used for managerial decision-making techniques and the application of specific mechanisms in choosing a development strategy from the identified alternatives.

2 Literature Review

2.1 Theory of Managerial Decision-Making

A person constantly faces a choice; every day we make a large number of decisions that affect our future to some extent. In the process of public administration, decisions acquire a great significance and importance, since it is not only the future of the government that depends on them, but also that of the entire economy of the country and millions of people. The ability to make the right decisions largely affects future success. Decision-making usually comes down to six steps (Fig. 1).

The process of decision-making is dependent on a large number of factors, which include conditions in the external environment, personal qualities of decision-makers (character, experience, gender, age, knowledge, and skills), government policy, priorities, financial status, etc. In this context, solutions can be programmed and not programmed, strategic and routine, organizational and personal, individual and group, tactical and operational, etc. Decision-making process uses rational, administrative, political models based on different approaches and opportunities extended to decision-makers.

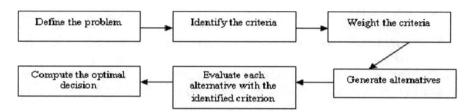

Fig. 1. Steps required for fulfilling the decision making process [9].

Having studied research on managerial decision-making, we come to the conclusion that basic principles are universal both at the level of individual organizations and at the regional and state levels. One can distinguish the following specifics of state decisions:

– Collectiveness;
– Significance;
– Dependence on political trends.

Collectiveness means that decisions are not made individually, but with the partic-ipation of a number of state authorities, taking into account the opinion of the public. Significance implies the scale of the impact and reach of the solution, as well as its importance for more individuals and organizations. At the same time, decisions made by the government or state authorities are based not only on the principle of rationality and efficiency, but also follow the country's current political trends.

As we know, decisions aimed at the medium and long term are made under the conditions of risk and uncertainty. In this context, some mathematical calculations can significantly ease the process of choosing alternatives, which we will demonstrate by the example of strategic decision-making in the field of tourism. James Nwoye identifies the following decision-making models:

– Probabilistic decision making model;
– Decision tree model;
– Game theory model [10].

In this study, the authors use probabilistic and mathematical decision-making tools, the methodological foundations of which are presented in the next section. As for decision-making strategies, they are as follows:

– Decision Strategy for Addressing Complex Problems;
– Decision Strategy for Addressing Well-Structured Problems;
– Bargaining as a Decision-making Strategy;

– Incremental or Trial-and-Error Strategies;
– Brainstorming Strategy;
– Nominal Grouping Strategy;
– Creative Thinking Strategy;
– Managing Emotions and Outbursts;
– Creative Vision [11].

From a methodological point of view, the authors chose a probabilistic decision-making model. At the same time, none of the above-mentioned strategies fully corresponds to the approach of decision-making strategy presented by the authors, therefore, for situations of risk and uncertainty, the authors suggest applying the strategy of the so-called "rational thinking", which is based on the analysis of the probability and effectiveness of each of the alternatives developed by the decision maker.

2.2 Current Features and Trends in Tourism Development

The outbreak of the Covid-19 virus has significantly affected all areas of human life. Tourism industry has been among those severely affected by the pandemic. Until 2019, the share of tourism in world GDP was almost 10%, the number of international arrivals was 1.5 billion, and about 320 million people were employed in this area [12]. Although the tourism development rate in 2019 decreased (2019, the growth rate of international arrivals was 4%, which is lower than in 2018 by 2%), however, the number of international arrivals was 1 460 million, and the total revenue from the tourism and transportation was almost 1.7 trillion USD [13]. From the point of view of export volumes in monetary terms, in 2019 tourism ranked third after the fuel and chemical industries [13]. In 2020, the situation changed dramatically, as the number of international tourist arrivals had decreased by 74% [14], which is unprecedented over the past decades. According to the results of a survey conducted by the World Tourism Organization (WTO) in May 2021, 45% of surveyed tourism professionals are confident that world tourism will regain the ground it lost in 2020 only in 2024 or even later [15]. The situation in particular countries is also unfavorable, in 2020 the decline in inbound tourism in Armeniaamounted to almost 80% [16], although the enterprises in the sphere of services and tourism tried to adapt to new realities and stimulate domestic tourism.

The Russian market of tourist services has also suffered significantly due to the pandemic and the closure of borders between countries: tour operators, travel agents, carriers and other participants of the tourism sector suffered serious losses. According to experts, only in the first months of the pandemic, the demand in all outbound directions decreased by 20–25% and after the closure of borders by many countries fell almost to zero. The market recovered only by the end of the 3rd quarter of 2020 [17].

Obviously under the present conditions one should not expect an active development of inbound and outbound tourism. Despite the gradual resumption of air communication with other countries, post-coronavirus measures in the field of tourism have been imposed almost everywhere, implying mandatory 14-day quarantine for inbound tourists. Therefore, considering the usual duration of Russian tours to foreign countries, not exceeding 10–14 days, and foreign tour operators do not expect a large flow of tourists from Russia. At the same time, promoting domestic tourism poses difficulties, such as problems with logistics, new sanitary and epidemiological requirements in organizing recreation, an increase in the price of tours. From the point of view of the initial conditions for the development of the tourism sector in Armenia and in Russia, the main difference, apart from the size of inbound and domestic tourism, is that in Armenia almost 87% of tourism expenditures are claimed in inbound tourism, while in Russia this figure is only 28.1%. Due to the fact that domestic tourism is more developed in Russia, the number of losses caused by the pandemic was lower than in countries similar to Armenia, as they largely depend on international tourism [18].

3 Research Methodology

Making decisions aimed at the medium and long term implies using a different set of methods and tools. Hence, in order to assess development strategies under the conditions of risk, it is first necessary to determine the possible situations of development of the external environment and assess their likelihood. To assess the likelihood of occurrence of certain phenomena in the external environment, as well as on the basis of assessments of the efficiency of each of the proposed strategies, the average survey results were calculated among the selected experts in the field of tourism of the Republic of Armenia (RA) and the Russian Federation (RF). Based on the obtained estimates of the probability of four possible situations in the external environment, as well as the forecast of the effectiveness of each proposed strategy, through the application of the extrapolation method, a matrix of solutions was built for possible strategies of tourism stimulation after the Covid-19 pandemic. Following this, calculations were made about the estimated costs of each strategy, as well as the standard deviation and the coefficient of variation, as indicators of the risk level for each strategy. Quantitative tools are presented below (Table 1).

Table 1. Solution matrix.

	N_1	N_2	N_3	N_4
S_1	X_{11}	X_{12}	X_{13}	X_{14}
S_2	X_{21}	X_{22}	X_{23}	X_{24}
S_3	X_{31}	X_{32}	X_{33}	X_{34}
S_4	X_{41}	X_{42}	X_{43}	X_{44}

N_1, N_2, N_3, N_4 – respectively in the context of the external environment:

- N_1- full recovery of the tourism sector within the next 5 years, the probability of this situation is P1;
- N_2 - partial, recovery of the tourism sector in the next 5 years, amounting to 80% of the volume of tourist flow of 2019, the probability of this situation is P2;
- N_3 - partial recovery of the tourism sector in the next 5 years, amounting to 60% of the volume of tourist flow in 2019, the probability of this situation is P3;

- N_4 - preservation of the current situation of recession and crisis, the likelihood of this situation is P4.

S_1, S_2, S_3, S_4 – are strategies for the further development of tourism in Armenia and Russia:

- S_1 - Strategy for forming and developing a tourism cluster as a factor contributing to tourism recovery;
- S_2 - Strategy of extension of the current system and policy of tourism development in the country, as a factor contributing to tourism recovery;
- S_3 - Strategy for starting and developing destination management organizations in the regions of country, as a factor contributing to the restoration of tourism;
- S_4 - Strategy for starting and developing the Ministry of Tourism in the country, as a factor contributing to the restoration of tourism.

X_{ij} reflect the corresponding expected results of a definite strategy (i) in a situation of external environment (j).

In this paper, the X_{ij} represents the total (inbound and domestic) tourists' expenditures in Armeniaand Russia. World Tourism and Travel Councils (WTTC) data for 2019 are selected as a basis for expert forecasts [18], which corresponds to the full recovery of the tourism industry in the post-pandemic period.

Initially, in order to determine the most appropriate strategy option, we need to calculate the estimated cost of each strategy using the mathematical expectation formula:

$$E(x_i) = P_1X_1 + P_2X_2 + \ldots + P_nX_n = \sum_{i=1}^{n} P_nX_n \tag{1}$$

Where,

$E(x_i)$ – is the estimated cost of each strategy (i).

Next, the standard deviation is calculated for each strategy in order to assess their riskiness using the following formula (2):

$$\sigma_i = \sqrt[2]{\sum_{i=1}^{n} (X_i - E(X))^2 P_i} \tag{2}$$

Where,

σ_i - is the standard deviation, i.e. the risk level for the strategy (i).

Finally, the coefficient of variation is calculated, in order to assess the risk ratio to the size of the estimated cost of the definite strategy, according to the following formula (3):

$$C_i = \frac{\sigma_i}{E(X_i)} 100\% \tag{3}$$

Where,

C_i - coefficient of variation for the strategy (i).

In contrast to the decision-making process for medium term perspective under the conditions of risk, we use a different approach for strategy evaluation in the situation of uncertainty for longer terms.

Uncertainty can be compared to a state in which one or more alternatives lead to possible and evaluable results, but the probability of situations of external environment is unknown. This happens when there are no reliable data from which the probabilities could be calculated a posteriori. Therefore, the process of making decisions under uncertainty is always subjective. It should be noted that uncertainty refers to the long term, for which the probability of a particular situation cannot be determined.

For the assessment of possible long term strategies of tourism development in the RA and RF, i.e. after 5 years or more, the authors will use the Wald, Savage, Hurwitz and Laplace criteria [19]. The content of these criteria for a long-term strategy evaluation is formulated onwards. Wald's Criterion is one of the methods for making decisions under the conditions of complete uncertainty. Wald's criterion is extremely conservative and, one might say, pessimistic, since it involves choosing the minimum return for each strategy, and then, of all the minimum outcomes, the maximum result [20]. This criterion attracts those who make cautious decisions, seeking assurances that, in the event of the most unfavorable outcome, there is at least a known payoff which is the biggest one among minimal outcomes. This approach may be justified because minimum payments may have a higher likelihood of occurrence.

There is also the maximax criterion, which is rarely used, due to the fact that it is optimistic and based on the principle of choosing the maximum outcome of each strategy, and then choosing the maximum return from all the maximums. Taking into account the specifics of the tourism sector and the degree of uncertainty, we will not use this approach in this research.

The next one is Savage's approach which is based on designing a loss matrix using the data from the original decision matrix. Then, from the loss matrix for each strategy, we select the maximum losses, and from them, the minimum values [21].

A quite common method for evaluating alternatives under the conditions of uncertainty is the Hurwitz criterion, proposed by Leonid Hurwitz in 1951. According to this approach, it is necessary to choose the minimum and maximum returns for each strategy and, having determined the weight α in the range from 0 to 1, evaluate each strategy. Instead of being optimistic or pessimistic, the Hurwitz criterion tries to get an average score for each strategy and ultimately choose the one with the highest score, if X_{ij} refers to positive-flow payoffs [22]. Evaluation of strategies according to the Hurwitz criterion is carried out according to the formulas (4): and (5):

$$H(A_i) = \alpha(\text{row maximum}) + (1 - \alpha)\ (\text{row minimum}) - \text{for positive-flow payoffs (profits, revenues)}$$
$$(4)$$

$$H(A_i) = \alpha(\text{row minimum}) + (1 - \alpha)\ (\text{row maximum}) - \text{for negative-flow payoffs (costs, losses)}$$
$$(5)$$

Where,

A_i – is an alternative strategy (i);

α – is a weight from 0 to 1 determined by the decision maker (usually $\alpha = 0.2$ or 0.3).

The Bayes-Laplace criterion (Laplace Criterion) suggests that if the outcomes in the external environment are known, but the prospect is so distant that it is impossible to estimate the probability of occurrence of these outcomes, then it can be assumed that all outcomes have equal probability and the assessment of strategies is done according to the Bayes-Laplace criterion, i.e. using the standard deviation and coefficient of variation [23].

The present research includes a survey administered among the experts in tourism in Armenia and Russia to determine the probability of certain situations in the future, in particular, the likelihood of restoring a pre-pandemic situation in the field of inbound and domestic tourism, as well as to assess the effectiveness of various tourism development strategies. For this purpose, 20 experts from Armenia and Russia were interviewed.

4   Research Results

As a result of the analysis of the expert survey, a matrix of decisions was built for the RA and the RF (Tables 2 and 3), representing possible strategies for the development of tourism, as well as the likelihood of recovering the volume of inbound and domestic tourism in the medium term, i.e. for the next 5 years. The X_{ij} reflects inbound and domestic tourist expenditure evaluations for Armenia in million US dollars, and for Russia in billions of US dollars (taking into account the estimates of experts from both countries).

Table 2. Solution matrix for Armenia.

	N1 (p = 0.09)	N_2 (p = 0.3)	N_3 (p = 0.6)	N_4 (p = 0.01)
S_1	3762.44	3883.25	2257.42	850.74
S_2	2800.45	1368.20	1026.10	386.70
S_3	3586.56	1970.21	1477.58	556.85
S_4	2950.52	1806.02	1354.45	345.56

N_1, N_2, N_3 and N_4 (from Table 2) correspond to the following situations for Armenia:

- N_1 - Full recovery of the tourism sector within the next 5 years, the probability of this situation is −0.09;
- N_2 - Partial recovery of the tourism sector in the next 5 years, amounting to 80% of the volume of tourist flow in 2019, the probability of this situation is 0.3;
- N_3 - Partial recovery of the tourism sector in the next 5 years, amounting to 60% of the volume of tourist flow in 2019, the probability of this situation is 0.6;
- N_4 - Preservation of the current situation of recession and crisis, the probability of this situation is 0.01.

Situations N_1, N_2, N_3 and N_4 (from Table 3) correspond to the following situations for Russia:

Table 3. Solution matrix for Russia

	N_1 (p = 0.1)	N_2 (p = 0.3)	N_3 (p = 0.5)	N_4 (p = 0.1)
S_1	49.50	39.63	29.70	29.41
S_2	52.70	42.10	31.60	28.10
S_3	45.12	28.25	21.18	17.29
S_4	42.52	22.42	16.82	13.72

- N_1 - Full recovery of the tourism sector within the next 5 years, the probability of this situation is 0.1;
- N_2 - Partial recovery of the tourism sector in the next 5 years, amounting to 80% of the volume of the tourist flow in 2019, the probability of this situation is 0.3;
- N_3 - Partial recovery of the tourism sector in the next 5 years, amounting to 60% of the volume of the tourist flow of 2019, the probability of this situation is 0.5;
- N_4 - Preservation of the current situation of recession and crisis, the probability of this situation is –0.1.

Further, S_1, S_2, S_3, S_4 are the corresponding strategies for the further development of tourism in Armenia and in Russia, as described in detail in the methodology of the article.

Tables 4 and 5, respectively, present the results of calculating the standard deviation and the coefficient of variation for the four proposed alternative strategies for Armenia and Russia.

Thus, comparing the results of the study for Armenia and Russia, we come to the conclusion that the experts from both countries almost equally (with a slight deviation) distribute the probability of tourism recovery after the pandemic, i.e. the experts determine the greatest probability for tourism recovery amounting to 60% of the 2019 volume within the next 5 years. At the same time, if in Armenia the current system of coordination and development of tourism is the least effective from the point of view of experts, experts from Russia consider it as the most effective among the ones listed, slightly outstripping the strategy of clustering in tourism. At the same time, cluster strategy is considered to be the most effective strategy for tourism in Armenia. The standard deviation and the coefficient of variation were used for selecting the best strategy for the further development of tourism in Armenia and Russia, as well as to assess the risk in the medium term.

On the one hand, Table 4 data suggest that the best choice for Armenia in terms of the estimated cost is the cluster strategy in tourism, but this strategy is also the riskiest in terms of the standard deviation. At the same time, if we take into account the risk per dollar of the perceived value, the lowest risk stands for tourism cluster strategy.

The analysis of Table 5 data suggests that the experts believe the strategy of the current system and the policy of tourism development to be most effective, however, the calculations of the standard deviation and the coefficient of variation show that in terms of risk, the cluster strategy, which received almost the same points from experts as the

Table 4. Solution matrix for Armenia (with strategy evaluation results).

	N1 (p = 0.1)	N2 (p = 0.3)	N3 (p = 0.5)	N4 (p = 0.1)	$E(X_i)$	σ_i	C_i
S_1	3762.44	3883.25	2257.42	850.74	**2866.55**	803.49	**0.28**
S2	2800.45	1368.20	1026.10	386.70	1282.03	507.00	0.40
S3	3586.56	1970.21	1477.58	556.85	1805.97	611.35	0.34
S4	2950.52	1806.02	1354.45	345.56	1623.48	**477.75**	0.29

Table 5. Solution matrix for Russia (with strategy evaluation results).

	N1 (p = 0.1)	N2 (p = 0.3)	N3 (p = 0.5)	N4 (p = 0.1)	$E(X_i)$	σ_i	C_i
S_1	49.50	39.63	29.70	29.41	34.63	**6.67**	**0.19**
S_2	65.87	42.10	31.60	28.10	**37.83**	10.63	0.28
S_3	45.12	28.25	21.18	17.29	25.31	7.54	0.30
S_4	42.52	22.42	16.82	13.72	20.76	7.80	0.38

current state policy, is more acceptable. Perhaps the reason for this distribution of expert assessment is due to the fact that the Russian government has been taking serious steps to form tourism clusters for quite a long period of time. Besides, there have already been clusters formed in some regions.

Apparently, for a long-term perspective, expert estimates of probabilities are not used due to inexpediency, therefore, solution matrixes for both Armenia and Russia will have the same appearance as in Tables 2 and 3, but without probabilities. Using Wald, Savage, Hurwitz and Laplace criteria for evaluating possible strategies described in the methodology of this paper, we present the results for Armenia and Russia, respectively, in Tables 6 and 7.

Table 6. Solution matrix for decision-making under conditions of uncertainty for Armenia.

	Wald's max/min	Hurwitz α = 0.2	Savage min/max loses	Laplace p = 0.25 σ	C	Wald's	Savage	Hurwitz α = 0.2	Laplace p = 0.25
S_1	**850.74**	**1433.08**	**0**	1273.29	**0.45**	✓	✓	✓	✓
S_2	386.70	869.45	2515.05	**875.49**	0.60				
S_3	556.85	1162.79	1913.04	1075.87	0.54				
S_4	345.56	866.55	2077.22	938.08	0.56				

In order to choose a strategy for the long-term development of the tourism sector in Armenia, all four criteria were used for strategy evaluation. The results indicated that the first option (a strategy for the formation of tourism clusters) is the best choice for Armenia. At the same time, as the previous estimates indicate this strategy was the most acceptable for the medium term, too.

Table 7. Solution matrix for decision-making under the conditions of uncertainty for Russia.

	Wald's max/min	Hurwitz $\alpha = 2$	Savage min/max loses	Laplace $p = 0.25$		Wald's	Savage	Hurwitz $\alpha = 0.2$	Laplace $p = 0.25$
				σ	C				
S_1	26.41	**33.43**	16.37	**8.28**	**0.22**			✓	✓
S_2	**28.10**	33.02	**1.31**	9.62	0.25	✓	✓		
S_3	17.29	22.86	20.75	10.66	0.38				
S_4	13.72	19.48	23.35	11.21	0.47				

The analysis of the aforementioned four strategies for the further development of tourism sphere in Russia based on Wald, Savage, Hurwitz and Laplace criteria showed that the choice of a long-term strategy can be made in two alternative ways. In this respect, two strategies were identified equally effective in Russia in the long term. Thus, both the cluster strategy and the continuation of the current state policy in the field of tourism have similar positions for further development of tourism in a long-time perspective.

5 Conclusions

After analyzing research findings, we conclude that for Armenia, both in the medium and long term, the cluster approach in tourism is considered to be the most effective way of development, while for Russia the situation remains ambiguous. The evaluation based on the estimated cost in Russia, suggests the current system and policy as most effective, however, from the point of view of low risk, the cluster strategy is more acceptable. At the same time, in the long term, the estimates split equally, therefore, according to the Wald and Savage criteria, the current strategy should be preserved in the long term in Russia, while the Hurwitz and Laplace criteria show that the cluster strategy will be the most acceptable one.

At the same time, it is worth mentioning that the other two strategies were not sufficiently considered due to low ratings. In Armenia and Russia, the strategy for starting a tourism ministry or developing destination management organizations was rated relatively low (almost 1.2–1.5 times lower than the cluster strategy). However, the current strategy of the state in the field of tourism earned the highest estimates in Russia, and the lowest in Armenia.

Acknowledgment. E. N. Antamoshkina thanks for funding research work within the framework of the Academic Support Program proposed by the Volgograd State Agrarian University.

References

1. Mintzberg, H.: The Nature of Managerial Work. Harper & Row, New York (1973)
2. Mintzberg, H.: Simply Managing: What Manager Do - And Can Do Better Berrett-Koehler, San Francisco (2013)
3. Mintzberg, H.: The Manager's Job. Folklore and Fact Harvard Business Review, March-April (1990)
4. Brynjolfsson, E.: Information Technology and the "New Managerial Work". In: MIT Sloan School of Management Center for Coordination Science Sloan School Working Paper, pp. 3563–3593. https://www.academia.edu/2662754/Information_technology_and_thenew_managerial_work. Accessed 28 Sep 2021
5. Kanter, R.M.: The new managerial work. Harvard Bus. Rev. **66**, 85–92 (1989)
6. Balcerzyk, D.: The role of a leader in contemporary organizations. Eur. Res. Stud. J. **XXIV**(1), 226–240 (2021). https://doi.org/10.35808/ersj/1959
7. Kolzow, D.: Leading from within: Building Organizational Leadership Capacity. https://www.iedconline.org/clientuploads/Downloads/edrp/Leading_from_Within.pdf. Accessed 28 Sep 2021
8. Sio, K.K.: Managerial Economics. INFRA-M, Moscow (2000)
9. Bazerman, M.H., Moore, D.A.: Judgment in Managerial Decision Making, 7th edn. John Wiley & Sons Inc., USA (2009)
10. Nwoye, J.: Decision-making strategies (Chapter 6, p. 93). Contemporary Issues on Management in Organization. Spectrum Books Limited, Ibadan. http://eprints.covenantuniversity.edu.ng/10021/1/DECISION-MAKING%20STRATEGIES.pdf. Accessed 02 Oct 2021
11. Nwoye, J.: Essential issues for successful executive decision-making in the 21st century. Int. J. Soc. Sci. Manage. Dev. **7**(2), 1–10 (2016)
12. Behsudi, A.: Wish you were here: tourism-dependent economies are among those harmed the most by the pandemic. Finance and Development, IMF, December 2020. https://www.imf.org/external/pubs/ft/fandd/2020/12/pdf/impact-of-the-pandemic-on-tourism-behsudi.pdf. Accessed 01 Oct 2021
13. International Tourism Highlights. 2020 Edition, January 2021. https://doi.org/10.18111/9789284422456. Accessed 01 Oct 2021
14. 2020: Worst year in tourism history with 1 billion fewer international arrivals (2020). https://www.unwto.org/news/2020-worst-year-in-tourism-history-with-1-billion-fewer-international-arrivals. Accessed 01 Oct 2021
15. International travel largely on hold despite uptick in May. https://www.unwto.org/taxonomy/term/347. Accessed 01 Oct 2021
16. Indicators. Number of Tourists, person (2020). https://mineconomy.am/en. Accessed 1 Oct 2021
17. Antamoshkina, E., Korabelnikov, I., Daeva, T., Nazarova, T., Morozova, N.: Methodological approach to the assessment of ecological tourism as a direction of sustainable development of the tourism industry. E3S Web Conf. **296**, 05006 (2021). https://doi.org/10.1051/e3sconf/202129605006
18. World Travel and Tourism Council. https://wttc.org/Research/Economic-Impact. Accessed 27 Oct 2021
19. Rozman, Č., Škraba, A., Kljajic, M., Pažek, K., Bavec, M., Bavec, F.: The development of an organic agriculture model: a system dynamics approach to support decision-making processes. Int. J. Decis. Support. Syst. Technol. **1**(3), 46–57 (2009). https://doi.org/10.4018/jdsst.2009070103
20. Wu, Z., Lupien, D.S., Georges, A.-N.: Decision making under strict uncertainty: case study in sewer network planning. World Acad. Sci. Eng. Technol. Int. J. Comput. Inf. Eng. **11**(7), 1–8 (2017)

21. Savage, L.: The theory of statistical decision. J. Am. Stat. Assoc. **46**, 55–67 (1951)
22. Hurwicz, L.: The generalized Bayes minimax principle: a criterion for decision making under uncertainty. In: Cowles Commission Discussion Paper, Statistics, p. 335 (1951)
23. Ulansky, V., Raza, A.: Generalization of minimax and maximin criteria in a game against nature for the case of a partial a priori uncertainty. Heliyon. **7**, e07498 (2021)

Implementation of Digital Design Processes Using Additive Technical Solutions: Case Study of Manufacturing Company

Olga Shvetsova$^{(\boxtimes)}$

Korea University of Technology and Education, Cheonan 31254, South Korea
shvetsova@koreatech.ac.kr

Abstract. This paper presents recent survey of digital manufacturing trends in Russian market, especially in the field of digital design and additive manufacturing technologies. The scope of research is focused on 20 Russian manufacturing companies from different market segments. Author investigates the main problems of Russian digital manufacturing markets and mentions its perspectives. This survey is based on such research methods as statistical analysis, case study and interview process. The period of survey is 2018–2021. Additive manufacturing technology has made a significant leap forward thanks to rapid improvement of electronic computing technology and software provision, so author makes a survey of functional business areas of digital technologies and software which are used in digital manufacturing process. This survey is based on case study of manufacturing companies in Russian market, which is focused on additive manufacturing technologies. Companies offers an extensive portfolio of software for digital manufacturing (D&M). They cover the complete production life cycle products, including sketching, prototyping, documentation development, engineering calculations, digital tests, preparation of presentation materials, preparation of production, organization of storage and exchange of information.

Keywords: Digital design · Advanced solution · Manufacturing company · Additive technology

1 Introduction

Additive manufacturing technologies make it possible to manufacture any product layer by layer based on a computer 3D model. Due to the gradual production of such the process of creating an object is also called "layering". If at in traditional production, we first have a blank, from which we then cut all unnecessary or we deform it, then in the case of additive technologies from nothing (or rather, from an amorphous consumable) a new product is built. Depending on the technology, the object can be built from the bottom up or, conversely, get various properties [1].

A. Gibadullin (Ed.): DITEM 2021, LNNS 432, pp. 130–144, 2022.
https://doi.org/10.1007/978-3-030-97730-6_12

Benefits of additive technologies:

- Improved properties of finished products. Thanks to layered construction products have a unique set of properties. For example, parts created on metal 3D printer, by its mechanical behavior, density, residual stress and other properties in some cases exceed analogs obtained by casting or machining;
- Great savings in raw materials. Additive technologies use practically the same the amount of material you need to manufacture your product. Then as with traditional manufacturing methods, the loss of raw materials can be up to 80–85%;
- Possibility of manufacturing products with complex geometry. Equipment for additive technologies allows the production of items that cannot be obtained in any other way. For example, a part inside a part. Or very sophisticated cooling systems based on mesh structures (no casting, not to make them by stamping);
- Mobility of production and acceleration of data exchange. No more drawings, measurements and bulky samples. Additive manufacturing technologies make it possible to manufacture any product layer by layer based on a computer 3D model. Due to the gradual production of such the process of creating an object is also called "layering". If at in traditional production, we first have a blank, from which we then cut all unnecessary or we deform it, then in the case of additive technologies from nothing (or rather, from an amorphous consumable) a new product is built.
- Depending on the technology, the object can be built from the bottom up or, conversely, get various properties.
- A computer model of the future product, which can be done in a matter of minutes transfer to the other end of the world - and immediately start production.

Additive technologies are preferred where weight reduction is required. Weight reduction is a constant key requirement of the aerospace industry, caused by both fuel costs and carbon dioxide emissions. That the same problem is relevant for the automotive industry, including commercial transport. Along with reduced fuel consumption and environmental friendliness, additive production has a number of other advantages. So, it significantly reduces the number of manufacturing stages, less material is consumed. All this is happening thanks to the design that requires less material and the process itself production. As technology improves, the capabilities of additive production are growing [2].

With all the many benefits of additive manufacturing, it has and disadvantages. First of all, it is low accuracy and high surface roughness. The way out of this situation is to use a hybrid approach, when on one on the same machine, both layer-by-layer synthesis (3D printing) and mechanical cutting processing. The object is created using additive technology, after which by cutting small elements are made. This approach also solves the problem of tolerances and accuracy, allowing for finishing and subsequent inspection measurements of the finished part.

Functions for both adding and removing material with small steps provide the ability to make changes to the design "on the fly", without rebuilding it from scratch. Due to this, the stages of calculations and tests are carried out much faster and with much more high precision. This approach also simplifies maintenance tasks and repair [3].

Additive manufacturing technology has made a significant leap forward thanks to rapid improvement of electronic computing technology and software provision. The current market for additive manufacturing is about 7 billion dollars 36, including production of special equipment and provision of services in a ratio of approximately 1/1 [4]. The share of Russia among countries actively developing and using additive manufacturing technologies, is approximately 1.2% (USA - 39.1%, Japan - 12.2%, Germany - 8.0%, China - 7.7%), and shows steady growth [5].

2 Materials and Methods

2.1 Materials

This survey is based on case study of manufacturing companies in Russian market, which is focused on additive manufacturing technologies. Companies offers an extensive portfolio of software for digital manufacturing (D&M). They cover the complete production life cycle products, including sketching, prototyping, documentation development, engineering calculations, digital tests, preparation of presentation materials, preparation of production, organization of storage and exchange of information.

Author takes a closer look at the use of which tools and at what stages ensures maximum efficiency. An integrated approach to implementation and the use of these systems allows for the transition to the technology of creating digital twin and its further application throughout the entire life cycle. Companies' solutions have performed well in both traditional methods of design and production, and in the transition to use modern and promising technologies - such as, for example, additive production, generative design [6].

Companies' portfolio of digital manufacturing software can be divided into several functional blocks (Table 1).

The companies' software portfolio covers a wide range of tasks, associated with the transition to digital production and the use of the most progressive technologies for creating digital twins. And since their solutions cover not only engineering tasks, but also related areas (for example, architectural and construction direction), there is the opportunity in a single information environment to get all the necessary information on the projects [7].

Table 1. Digital manufacturing software functional blocks and characteristics.

Functional block	Digital software	Characteristics
Industrial design	• 3ds Max - 3D modeling, animation and visualization; • Alias - work with surfaces, including class A; • Autodesk Rendering - fast high quality rendering in the cloud; • Fusion 360 - the next generation of cloud-based 3D software for complex processes development and industrial production of products; • VRED - 3D visualization and virtual prototyping for automotive designers	Stage of life cycle program "Research and design", stage KCCI "Development of the concept of EMR/ESI (SDI
Digital design	• AutoCAD - CAD for 2D and 3D design; • Inventor - 3D-CAD for mechanical design; • Fusion 360 - the next generation of cloud-based 3D software for complex processes development and industrial production of products; • Navisworks Manage - software for checking designs using 5D analysis; • Recap PRO - software and services for laser and 3D scanning; • Factory Design Utilities - conceptual design, planning and inspection of production facilities	Stage of life cycle program "Development", stage of KTPP "Development" EMR (KTR)
Analysis of products and structures	• Autodesk CFD - a tool for calculating flows of liquids and gases; • Helius Pfa - a powerful tool for analyzing the behavior of composite materials; • Inventor Nastran is an embedded Inventor solver based on the finite elements; • Moldflow - simulation of plastic injection molding; • Robot Structural Analysis Professional - structural analysis	Life cycle stage "Development", stage KTPP "Development of EMR (KTR)

(continued)

Table 1. (*continued*)

Functional block	Digital software	Characteristics
Preparation of production	Featurecam - software for programming CNC machines; • Hsmworks - high-speed processing organization, integrates with Solidworks; • Inventor CAM - an integrated solution for programming of various types processing (from 2.5 to 5-axis) for Inventor and Solidworks; • Netfabb - a complex for working with additive technologies (design and production); • Powermill - preparation of high speed and 5-axis machining; • Inventor Nesting - software for nesting real-world parts in Inventor for the purpose of optimization of work with flat workpieces; • TruNest - software with advanced production preparation functionality for cutting products from sheet materials; • TruComposites - add-in for Inventor: efficient process design • Production of products from composite materials	Stage of life cycle program "Manufacturing (production)", stage KTPP "Preparation of production"
Analysis of quality	Power Inspect software product designed to check complex surfaces (refers to hardware control)	Stage of life cycle program "Manufacturing (production)"
Production and operation of the product	Includes applied solutions based on the Autodesk Forge platform for managing using the product	Stage of life cycle program "Operation"
Engineering data management	"A" Vault (PDM-system, engineering data management)	Stage of life cycle program "Operation"

Source: made by author

2.2 Methods

This article is based on 2 types of research: survey and case study. The survey includes digital market statistical data and it has been applied to separate domains of digital manufacturing process in Russian small-size and medium-size companies. Author discusses digital industrial technologies for B2B market [8]. The period of survey is 2021. There are 20 small and medium size companies in Russian market were investigated. Description of survey participants is presented in Table 2.

Table 2. Characteristics of research participants.

Type of company	Business characteristics	Number of participants	Number of employees	Year of establishment
Healthcare	Fast food delivery service, online shops	5	50–100	1999, 2003, 2005, 2008
Heavy production	Fast food delivery service, menu production, cooking training, online shops	2	100–150	1966, 1995 1961, 1975
Military service	Fast food delivery service, online shops	2	100–500	2011–2019
Transportation	Fast food delivery service, passenger transportation, logistics, post service	11	150–200	1985, 1993, 1997, 2001, 2005, 2007

Case studies have a very narrow focus which results in detailed descriptive data which is unique to the case(s) studied (Fig. 1).

The first study presents statistical research of additive technologies in Russian market. The second study investigates the opinion of managers and employees of digital companies about perspectives of digital design. For the results interview process was implemented (Table 3).

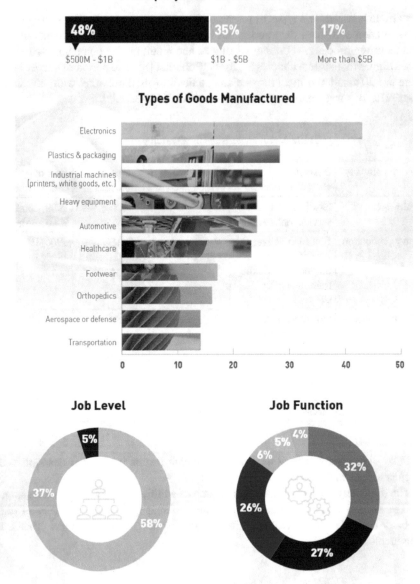

Fig. 1. Interview process – digital companies, 2020. Source: made by author

Table 3. Example of interview questions.

Type of personnel	Questions	Number of participants
Project managers	What is typical product for digital design; what are the main problems; what is the team	16
Top managers (CEO)	What is financial strategy; what is the main strategy	8
Engineers	What is the set of the materials; how many steps in NPD	32
Designers	What is the pool of ideas; what is the necessary characteristics of product	56
Other employees (marketing, etc., R&D)	What is market demand; how many prototypes are necessary	105

Source: made by author

3 Data Collection and Results

3.1 Digital Market and Additive Technologies in Russian Market: Statistical Review

Russian and foreign experts characterize the current development of the Russian market of additive technologies as a stage of formation in comparison with the world and stress its low maturity in general. There is a reason to believe that the reality may differ significantly from some traditional quantitative counting systems. In Russia we can already see manufacturers of materials and expensive industrial 3D-printers which cost more than 5 million rubles each, there is already a critical mass of developments, additive industry is included in the technological agenda of the state. The leading industrial centers of the country and state corporations have begun the transition from prototyping, production of equipment, studying the possibilities given by the technology and R&D, to repair of functional products and selective printing of final working products.

The analysis conducted by J'son & Partners Consulting allows us to conclude that there is no technological lag of the Russian Federation in the AM segment [9]. Some quality indicators of AM achievements in Russia (engines, turbines, buildings, materials…) have already put the country among the world leaders in the development of such technologies. But due to the implementational, investment and commercial conditions and still weak cooperation, civil industries lag far behind in terms of AP application and development of commercial services (3D printing on demand). J'son & Partners Consulting concludes that Russia's share in the global 3D printing market of 1.5% is not representative as an indicator of the state of this technology in a single country [10].

In Russia there has already appeared a strong ecosystem of players. Russia has mastered production of equipment and raw materials; Russia has begun to develop specific AP software. A strong and reliable team, able to provide all services on a "turnkey" basis in Russia has already formed. And they are already running a number of AM projects in large industrial sectors (Fig. 2).

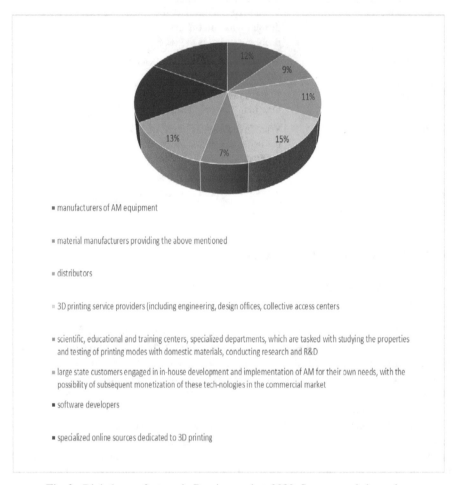

Fig. 2. Digital manufacturer in Russian market, 2020. Source: made by author

The major players' profiles are considered separately in the special sections of the Report "The 3D printing market and additive technology application prospects in Russia and in the world". However, there is still too strong dependence on imports and R&D funds for bringing the infrastructure to commercialization at the global level.

Key indicators of the Russian market (assessment by J'son & Partners Consulting).

– The 3D printing market in Russia is growing steadily: over the past 8 years, in quantitative terms, it has grown 10 times (according to J'son & Partners Consulting);
– In monetary terms, total sales of 3D printing equipment, materials and services (including R&D) rose to 4.5 billion rubles per year ($69 million, estimated in 2020);

- Purchase of 3D equipment, additional equipment and materials accounts for about 80% of the market. While globally, 60% are 3D printing services/engineering;
- Domestic printers account for about 30% of sales in the Russian market;
- A significant role in the structure of demand in the 3D printing market in Russia, as well as the development of research competencies is played by large corporate customers;
- In the structure of imports in quantitative terms about 90% are low-budget 3D printers, in value terms the maximum sales (up to 70%) are the most expensive machines;
- The Russian market of 3D printing equipment and materials is characterized by a high degree of competition between foreign players;
- Manufacturers from Germany and USA are the leaders in the rating of companies that supply equipment and materials for 3D printing. China closes the top three countries, the largest exporters of additive equipment and materials in Russia.

Leading positions on total imports to the Russian market in this period were held by the following companies: 3D systems, Concept lazer GMBH, Voxeljet AG, Stratasys, Envisiontec, Markforget, Formlabs, Prodway Group, Ultimaker, Hunan Farsoon Hightech CO, Mundo SL, Beijing Tier Time Technology CO.LTD, Wanhao, etc. [11].

The Russian AM market has a branch structure that repeats the global one – industry, aerospace and medicine are traditional leaders in 3D printing of functional products.

3.2   Perspectives of Additive Manufacturing and Design in Russian Market

According to Russian experts, now in the world in the field of industrial design a number of key trends that will be relevant over the next few years. Wherein the technological aspect plays a dominant role. In particular, there is an increase integration of industrial design with engineering. The direction associated with designing new materials for design. Customization is gradually taking place production due to the introduction of cost-effective advanced technologies. Due to the fast the development of the Internet of Things and the Industrial Internet (Industry 4.0) is popularized design of objects "within the network" [12, 13]. Such a direction as expertise is developing the impact of design solutions on the environment against the background of growing environmental problems in the country.

Hereby you can find the results of interview process (Figs. 3, 4 and 5).

At the moment, against the background of the active development of various technologies, there is a complication industrial product. Globalization processes also play a role. According to opinions from interview, at the moment in the field of industrial design on a global scale you can observe the strengthening of existing trends, which, due to the complexity of the directions are versatile:

- Gradual unification of existing styles and directions of industrial design and blurring distinctions between local design schools (significant impact globalization processes). In this regard, there is a gradual reducing the export barrier associated with national perceptions industrial products in foreign markets.
- Integration with engineering. Gradually the business stages associated with volumetric designing the shape of the product and forming its internal structure, turn into parallel

Fig. 3. Interview results, 2020. Source: made by author

processes. Creation of a multifunctional form at once reduces development time for a new model and accelerates time to market.
– The growing role of designers in the development of new materials for industrial production. At the moment, there is a fairly large list of materials with unique properties that were created precisely by designers and engineers for specific tasks, and not representatives of the chemical industry. In addition, in specialized secondary and higher educational institutions, where they teach designers, disciplines for students are actively introduced and improved, concerning the design of new materials. The most developed approach is generative design in manufacturing (Fig. 6).

Generative design features are currently available in software products like Netfabb Ultimate and Fusion 360. From the point of view of the user, the principle of operation of this design technology is pretty straightforward. The constructor sets the volume in which the system is allowed search for a solution, indicates zones inviolable for change, obstacles that need to be skirted, fasteners and work areas of the part, as well as the conditions for the functioning of the part are fixing and the loads acting on the part. Then the task is sent to the solution - to the company's cloud. While the cloud searches possible options, the designer and his computer are completely free and can engage in other creative tasks. As soon as the system finishes searching for solutions, it will notify the user about this and provide him with all the found part geometry options. The designer will only need to choose the most suitable - by weight, strength, permissible deformations, manufacturability and others criteria. Further, it remains only, if necessary, to modify the part in CAD and/or CNC solutions, analyze it in software calculation programs and prepare for production.

Research and development is now the most common 3D printing use case.

In what ways is your company currently using 3D printing?

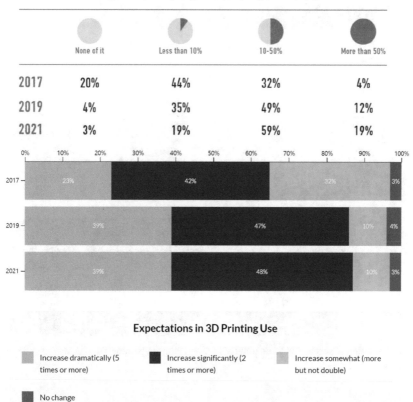

	Research and Development	Prototyping	Jigs, Fixtures and Toling	Bridge Production	Production Parts	Repair and Maintenance
2017	-	69%	30%	23%	27%	14%
2019	53%	66%	37%	39%	52%	38%
2021	73%	72%	57%	56%	62%	46%

Manufacturers are using 3D printing more frequently for producing production parts.

Approximate percentages of production parts produced with 3D printing

	None of it	Less than 10%	10-50%	More than 50%
2017	20%	44%	32%	4%
2019	4%	35%	49%	12%
2021	3%	19%	59%	19%

Expectations in 3D Printing Use

- Increase dramatically (5 times or more)
- Increase significantly (2 times or more)
- Increase somewhat (more but not double)
- No change

Fig. 4. How do you expect your company's use of 3D printing to change in the coming 2–5 years? Choose the answer that most closely applies. Interview results, 2021. Source: made by author

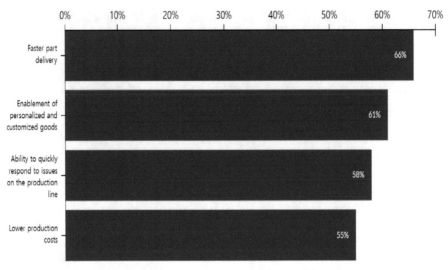

Fig. 5. What benefits do you expect to gain from mass adoption of 3D printing for manufacturing? Choose all that apply. Source: made by author

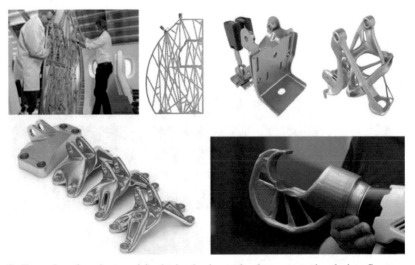

Fig. 6. Examples of product models obtained using technology generative design. Source: made by author.

4 Discussion

Nowadays in the market of services in the field of art and volumetric-functional the presence of rather diverse players is observed. Conditionally all participants can be divided into several groups: manufacturers themselves, design companies and freelance designers, fashion houses and advertising companies and research institutes and divisions It should be noted that in-house design is very popular a business model among manufacturers worldwide. Own design teams, often involving material engineers and technologists [14]. However, in this case there is a high possibility of burnout for designers and a change in the creative team is required. However less advantage of this approach is the design team that understands and accepts brand values. However, according to experts, the popularity of outsourcing services in the field of industrial design. Large and medium-sized manufacturers turn to external design companies to resolve issues related to industrial design of products. Such a decision can be explained by the desire of manufacturers to avoid the cost of creating and maintaining own design divisions, as well as get a new image of industrial products designed by professionals. It should be emphasized that the role of freelance designers in outsourcing is quite high. According to analysts, in developed markets, the share of it is freelancers who make up 10–20% on average [15].

There is an interesting fact is that the direct producers of various items design - fashion houses - famous industrial designers are often invited, as a rule, to develop a new unique and non-standard "fresh" design of the items. In addition, these fashion houses (and brands) can collaborate with each other (cross-promotion also plays a big role here). All this is also a kind of tools for promoting new products.

In addition, against the background of the strengthening of globalization processes in the world, it is gradually developing a new business model for this sector - industrial cooperation in the field industrial design and engineering design. Design and production functions are distributed between industrial product integrators and, respectively, by the suppliers of the modules. This approach can significantly reduce the time and the cost of launching a new product [16]. Number of design companies and independent the number of designers involved in such integrated projects is growing. So, according to Design Council, over the past few years in the UK about 60% of all market participants engaged in co-design. A similar trend is observed in other developed and emerging markets.

5   Conclusion

The industrial design of the manufactured product, as already mentioned, plays a critical role in the formation of demand for goods, both in the domestic and foreign markets. In Russia, the external the type of products of various industries cannot yet compete with Western (in design and quality) and oriental counterparts (in design and price). At the same time, different enterprises in the country are at different stages of understanding the importance of industrial design: from the beginning of investing to completely ignoring this direction. The main reasons for the rejection of industrial design in the country are backward technologies in production, the habit of domestic manufacturers to in-house design and skepticism in regarding the possibilities of industrial design as a powerful differentiating tool for consumer market. Also, the problem is complicated by the inability to obtain competitive education in the country, its isolation from production, real products and modern technologies.

Nowadays, the demand for services in the field of industrial design, according to market participants, is gradually growing, however, there are a number of barriers that prevent more active development directions. So, among such stop factors is the skeptical attitude of Russian manufacturers to the development of ergonomic and aesthetic design of industrial products. At the same time, manufacturers initially focused on consumer markets (B2C), are more flexible in this regard and begin to gradually realize the need development of a competitive design. Acceptance and implementation of industrial design in B2B companies, not accustomed to working in retail markets, it is much slower.

Nevertheless, the market for services in the field of industrial design is gradually developing. High growth rates compared to traditional industries (on average about 20% per year, according to experts) due to a low base. There are no real drivers for the explosive growth of the market. observed. However, technological state corporations together with the market leaders support the emerging creative segment by virtue of their capabilities.

References

1. Boyard, N.: Design for additive manufacturing - DFAM. J. Mater. Process. Technol. **12**, 616–622 (2014)
2. Hamidreza, S.G., Corker, J., Fan, M.: Additive manufacturing technology and its implementation in construction as an eco-innovative solution. Autom. Constr. **93**, 1–11 (2018)
3. Shvetsova, O.A., Lee, S.-K.: Living labs in university-industry cooperation as a part of innovation ecosystem: case study of South Korea. Sustain. J. **13**, 57–93 (2021)
4. OECD Homepage. https://www.oecd.org/economy/surveys/. Accessed 11 Nov 2021
5. Statista Homepage. https://www.statista.com/outlook/dmo/ecommerce/russia/. Accessed 15 Nov 2021
6. Moreno, N.D., Sánchez, M.D.: Design for additive manufacturing: tool review and a case study. Appl. Sci. **11**, 1571–1604 (2021)
7. Cohen, Y., Faccio, M., Pilati, F.: Design and management of digital manufacturing and assembly systems in the Industry 4.0 era. Int. J. Adv. Manuf. Technol. **105**, 3565–3577 (2019)
8. Shvetsova, O.A.: New Forms of Effective Collaboration: How to Enhance Big Data for Innovative Ideas in the Online Environment, Book "Big Data for Entrepreneurship and Sustainable Development", 1st edn., Routledge CRS Press (Francis&Taylor Group), USA (2021)
9. Euromonitor International homepage. https://www.euromonitor.com/. Accessed 12 Oct 2021
10. Juhanko, J., et al.: Digital design and manufacturing process comparison for new custom-made product family - a case study of a bathroom faucet. Est. J. Eng. **19**, 76–95 (2012)
11. Jin, Y., He, Y., Fu, J.-Z.: Support generation for additive manufacturing based on sliced data. Int. J. Adv. Manuf. Technol. **80**(9–12), 2041–2052 (2015). https://doi.org/10.1007/s00170-015-7190-3
12. Zwier, M.P., Wits, W.W.: Design for additive manufacturing: automated build orientation selection and optimization. Procedia CIRP **55**, 128–133 (2016)
13. Global Data homepage. https://www.globaldata.com/. Accessed 12 Oct 2021
14. Thomson, M.K., et al.: Design for additive manufacturing: trends, opportunities, considerations, and constraints. CIRP Ann. Manuf. Technol. **65**(2), 737–760 (2016)
15. Steele, K., Ahrentzen, S.: Design guidelines. Home Autism **11**(4), 81–170 (2017)
16. London, T.: Design for additive manufacturing. Addit. Manuf. Technol. **17**, 56–70 (2015)

Applying Affective Computing to Marketing Problems

Diana Bogdanova[1], Nafisa Yusupova[1], Italo Trevisan[2], and Andrea Molinari[2(✉)]

[1] Ufa State Aviation Technical University, 12, Karl Marx Street, Ufa 450008, Russia
[2] University of Trento, Via Inama, 5, 38122 Trento, Italy
amolinar@gmail.com

Abstract. The main marketing goals are related to attracting new customers, maintaining long-term relationships with current customers, and increasing the impact on customers. This can be achieved by increasing customer loyalty by considering their individual preferences, customer characteristics, and emotions during the service provision. To improve the efficiency of the service delivery process, it is necessary to support decision-making, based on tracking the level of customer satisfaction, taking into account their emotionally colored information when consuming services. The apparatus of affective computing allows to recognize, process, and account for human emotions. One of the applied technologies of affective computing is machine learning. This article proposes an approach to assessing customer satisfaction by analyzing customer reviews in natural language published on the Internet. It is proposed to use machine learning methods such as a Naive Bayes classifier, Bernoulli model, and support vector machines (SVM) to analyze customer reviews. Customer reviews from the site tophotels.ru were used for the analysis. For the Naive Bayes classifier, the best result was obtained by the multinomial model - 86.83%. Bernoulli's model reached 86.49%. Support vector machine (SVM) showed the best result on a linear kernel with accounting the presence/absence of a word in the document - 87.69%. The analysis of emotionally colored information from customer reviews for different aspect groups of services allows us to assess quantitatively and qualitatively the level of customer satisfaction and monitor the dynamics of this indicator.

Keywords: Affective computing · Marketing tasks · Machine learning · Customer satisfaction · Customer review analysis · Emotionally charged information

1 Introduction

A distinctive feature of the modern service market is that supply is much greater than demand. Each company strives to improve its competitiveness, achieve high-performance indicators, and conquer the largest possible market share. At the same time, when choosing services, clients strive to maximize the service utility, and, of course, they are guided by the ratio of the level of satisfaction and price. In a modern economy, if a company expects long-term relationships with the customers, then the basis of its

A. Gibadullin (Ed.): DITEM 2021, LNNS 432, pp. 145–158, 2022.
https://doi.org/10.1007/978-3-030-97730-6_13

competitive strategy is to achieve maximum customer satisfaction and comfort and their positive emotions in the process of receiving the service. To ensure the quality of the company's services, it is necessary to make effective management decisions. Consideration of management decisions and their adoption should be based on knowledge and patterns obtained from the analysis of the information collected about the process of providing services. To customize the process and adjust parameters of providing services, it is necessary to assess the level of customer satisfaction by taking into account their emotions while receiving services. It is necessary to develop an approach to assessing the level of customer satisfaction, taking into account the emotionally colored information coming from them, based on the technologies of affective computing.

The introduction substantiates the relevance of the research topic; in the first part, the results of the analysis of literature on the accounting of emotionally colored information in marketing tasks are presented; in the second - the formulation of the research problem is given; the third part examines the existing approaches to assessing the level of customer satisfaction; in the fourth, an approach to assessing the level of customer satisfaction by taking into account emotionally colored information is proposed; and in conclusion there are the results and conclusions.

2   State of the Art

Many scientists concern themselves with issues of customer emotions accounting when solving marketing problems of the companies. The article [1] provides an overview of the current theoretical appraisal of the role of emotions in marketing. and, It presents four main theories of emotions present in psychological literature and examples of their marketing applications. Moreover, it shows the results of five empirical works aiming to assess consumers' emotions in response to marketing. The characteristics of emotions in marketing are compared with the characteristics of emotions from a psychological point of view. The article ends with suggestions for the "development of the theory of emotions in marketing" articulated in three propositions related to consumers' range, intensity, and quality of emotions. In [2], there is a review of the research on emotions that was conducted in the field of marketing from 2002 to 2013. The article focuses on social/personality issues, cognitive factors, emotions' development, and their interaction with other consumption factors. The authors reviewed a total of 340 emotion-related articles published in 19 marketing journals.

In [3], the authors propose integrating all existing studies of emotions caused by marketing incentives, products, and brands into a hierarchical model of consumer emotions. The paper addresses the issue of significant discrepancies in the content and structure of emotions in the presented studies. In addition, the authors pay attention to the types of emotions and their impact on customer behavior.

In [4], the authors examine the relationship between the work of marketers and consumer behavior in the context of the evolution of the COVID-19 pandemic. Special attention is paid to studying the influence of the emotions of the concerned consumer associated with the epidemic. According to the researchers, consumer anxiety often triggers strong emotions that tangibly affect the buying process. The authors advise marketers to consider the broader impact of emotion, especially as a motivation to reduce uncertainty, on a consumer's dynamic decision-making process.

The article [5] presents a study based on tourists' emotions on vacation and other tourist services. The study was conducted with the participation of 400 tourists who visited the attractions. The authors took as a basis a two-dimensional approach to the study of emotions. The results confirm the relationship between emotions and behavior - those tourists who showed increased satisfaction levels showed loyalty and willingness to pay more for services. The authors recommend service companies to incorporate affective techniques into their marketing work.

[6] combines existing research related to marketing, customer emotions, customer satisfaction, and loyalty to a particular retail chain. The researchers surveyed 274 customers in four stores in one major coffee house chain. Using this data, the authors test hypotheses and models by modeling structural equations.

The article [7] examines modern research that questions the traditional assumptions about the role of emotions in consumer decision-making and introduces the role of consumer emotional intelligence in this process. The paper concludes with a look at the strategic implications for marketers of abandoning traditional patterns of interaction with consumer emotions and embracing new technologies.

In [8], the authors consider the tasks of developing and implementing marketing strategies that will have a positive emotional impact on the audience. The work is based on the study of emotional branding (or emotional marketing) - those techniques used by companies to directly address the emotional state, needs, desires, and beliefs of consumers and interact with them.

Nowadays, customers are satisfying their high order needs in the purchasing process, thus the role of emotions and, therefore, the visual design of product packaging in the purchasing decision process has increased. At the same time, the collaborative analysis assumes that consumers are rational in their decisions and maximize their utility functions. In the article [9], the authors combined the method of joint analysis with psycho-physiological measurements. The results show that the combined method can determine how significant different visual effects of product packaging are in creating positive customer experiences.

In [10], the authors presented their view of how emotions – related and unrelated to decision-making - play an important role in consumers' decision-making process. The article examines why and under what conditions emotions are the precursors of decision-making and plans further research to study the emotional impact on the consumer's decision-making process.

The article [11] examines the influence of emotions on customer satisfaction and quality of service. The results indicate that the measurement of emotion had a consistent direct effect on all dependent variables. The authors concluded that different aspects of emotion affect satisfaction at different stages of service delivery. The theoretical and managerial implications of the results are discussed in detail.

The goal of the article [12] is to explore the role of emotions in the purchasing behavior of a company. The authors interviewed marketing decision-makers for a major brand. Taking into account the perspectives of these marketing professionals, the authors examine the role that emotions play in consumer behavior. The study complements existing knowledge by offering a new understanding of which discrete or specific emotions are most prominent in an organization's buying behavior and how their expression influences decision-making at each buying cycle stage.

The article [13] reviews studies that state that most purchasing decisions result from careful analysis of advantages and disadvantages, as well as affective and emotional aspects. Emotional states of people always influence every stage of their decision-making in the buying process. The authors developed an original algorithm for analyzing customer sentiment based on their online reviews. With the help of this algorithm, they got good results in polarizing this opinion to achieve strategic marketing goals.

The study [14] empirically assesses specific questions about whether the use of knowledge about various forms of emotions and their application in traditional models contributes to understanding the formation of loyalty intentions in the context of retail marketing. The managerial and research implications of the results obtained are discussed.

Research [15] aims to fill a gap in the research of emotions in tourism and hospitality and offers a theoretical basis for explaining affect-driven behavior. Research results support the theoretical basis of affect-based consumer behavior, suggesting that identifying positive emotions has a beneficial effect on consumer behavior in pursuit of goals and novelty. The authors state that deepening the understanding of consumer emotions is critical for tourism and hospitality marketers.

In the article [16], the authors consider the role of emotions in human behavior in the field of advertising, study the influence of advertising literature. The results indicate that the potential for direct behavioral effects of discrete integral emotions (specific emotions triggered by an ad message) remains underestimated. The authors outline the theoretical rationale for this behavioral approach and outline challenges and opportunities for considering the potential of discrete emotions and their advertising-related behavioral outcomes. Researchers provide guidelines for applying this rationale to advertising research.

3   Problem Statement

In a highly competitive fight for market share, the company's main task is the fullest satisfaction of the needs of its customers. A service sector company that wants to improve the efficiency of its activities faces two questions:

– What is the satisfaction of customers, what emotions they experience in the process of consumption;
– How to increase the satisfaction of services, how to take into account the emotionally colored information of clients in the process of providing services?

In this regard, methods of collecting and analyzing information about customer satisfaction, about emotions experienced by customers in the process of consumption, which will determine the effectiveness of decisions made to improve the quality of services, acquire an important role.

There is a need to automate the existing methods of customer satisfaction research. Automation of certain stages of the existing classical data collection methods, such as drafting a questionnaire, conducting interviews, making calls to customers, is impossible. The complexity of integrating the processes of preparing research, collecting and analyzing data, developing recommendations for a decision-maker prevents the formalization of the entire research procedure and, as a result, makes it impossible to develop a unified information system.

Separately, it should be noted that there is no methodology for developing and supporting management decisions to improve the quality of services, which would be based on assessing the level of customer satisfaction that considers customers' emotions. This fact is also an obstacle to the development of specialized software for the quality management of services.

The described problems help to formulate the basic requirements for an information system for quality management of services based on an assessment of the level of customer satisfaction, taking into account emotionally colored information:

- The ability to conduct continuous monitoring of emotionally colored information and the level of customer satisfaction;
- Minimal human participation, to involve the specialists only at the stage of setting up the system;
- Automatic collection and processing of unstructured and semi-structured data;
- The ability to make recommendations for improving the quality of services for a decision-maker based on a set of customer satisfaction data, taking into account emotionally colored information.

A fundamentally new approach is required to overcome methodological barriers and develop an information system for managing the quality of services that meets the requirements described above. The most promising avenue is the idea of using the Internet as a source of consumer opinions and their processing using artificial intelligence technologies and affective computing. In this connection, the following task is state.

Let U be the customer satisfaction function. Then the function:

$$U = U(M) \tag{1}$$

Where M is the opinion of customers about the service provided.

Customer opinion is formed from the correspondence of the provided service to their requirements.

It is necessary to evaluate the customer opinion function:

$$M = M\,(I,\ E,\ S,\ H) \tag{2}$$

Which characterizes the compliance of the services provided to the customers' requirements.

Where I, E, S, H are customer requirements regarding different aspects of the services provided. These requirements are a set of indicators characterizing the services provided.

Function M can be assessed by analyzing customer feedback and their emotions regarding various aspects of the services provided.

4   Analysis of Methods for Assessing Customer Satisfaction

In accordance with ISO 9001, an organization should monitor information related to the client's perception of how the company fulfills its requirements as one way to measure the quality management system [17]. Manufacturers use this information to improve the quality of services, processes, and company characteristics that matter to the client to retain regular customers and attract new ones. Figure 1 presents a diagram of customer satisfaction [18], according to which an organization should strive to bridge the gap between the expected quality of service and the quality of the services received, both in the perception of the customer. Consumer dissatisfaction with the quality of products leads to loss of reputation and the departure of customers to competing firms.

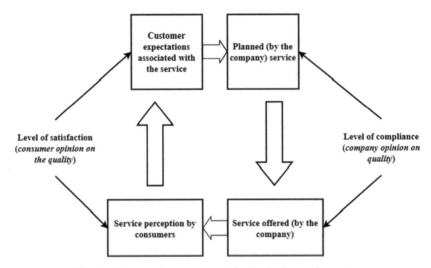

Fig. 1. Diagram of customer satisfaction in the service sector.

One source of information on service perception is a customer satisfaction survey. In the ISO 9000 quality management system standard model, the customer satisfaction survey process is implemented in the "Measurement, analysis and improvement" block through the feedback from the "Consumers" block. Since customer satisfaction information is essential for effective management decisions, businesses should develop and implement processes for collecting, measuring, and analyzing satisfaction data on a regular basis.

International quality standard ISO 10004 contains recommendations for the implementation of these processes in an organization. It also defines the main objectives of satisfaction research, including assessing opinions about services and their specific aspects, researching the reasons for losing market share, identifying trends in opinions, and analyzing customer satisfaction of competing organizations. Satisfaction surveys are categorized as quantitative and qualitative. Qualitative research is carried out to identify specific aspects of services and their significance to clients in meeting its requirements, contributing to the understanding of individual perceptions and the disclosure of ideas and issues. Quantitative research is conducted to measure customer satisfaction, assess the current situation, and identify trends in customer satisfaction. ISO 10004 recommends using the following methods to collect information on customer satisfaction (see Table 1) [18]:

1. Personal interviews - a research method in which the interviewer asks questions to the respondent and receives answers during their direct communication;
2. Telephone interview - a research method in which the interviewer asks questions to the respondent and receives answers via telephone;
3. Discussion groups - a type of group research methods in which the interviewer leads an active discussion and exchange of views with a group of five to ten respondents;
4. Correspondence research - a method of questionnaire research, in which the respondent answers questions by mailing out questionnaires;
5. On-line survey - a method of questionnaire research, in which the respondent answers the questions of the questionnaire through the Internet mailing list.

However, the methods of collecting and analyzing customer opinions proposed by the standard have several significant deficiencies that complicate their practical use as tools for researching customer satisfaction in the service sector.

Methods (1), (2) and (3) require direct interaction with clients or experts. Assuming that the average interview duration is 20–30 min, the research can take several weeks. The cost of a study increases with its duration, so these methods are the most laborious, time-consuming, and expensive. In addition, in methods (1) and (2) there is a risk of information distortion on the part of the respondent.

Methods (4) and (5) require significantly fewer research costs, but they also have deficiencies. The number of answered respondents can be minimal and statistically insignificant. For this reason, there is a risk of disruption to the study. Respondents can perceive the questions of the questionnaire and the rating scales in different ways, which leads to a loss of objectivity in the study. An incorrectly prepared questionnaire can lead to results opposite to the actual situation. Therefore, the involvement of a specialist is required to develop them. To apply methods (4) and (5), it is necessary to have a base of postal addresses or e-mail addresses for mailing, and it also takes time to send and collect questionnaires from respondents.

A common disadvantage of the methods presented in the ISO 10004 standard is the need to perform a large amount of "manual" work at all stages of the conducted customer satisfaction survey: preparing questionnaire questions, selecting a respondent base, sending a questionnaire and collecting results, conducting personal interviews, preparing a report on the results. This disadvantage leads to an increase in the cost of research, not to mention the conduct of such research in relation to their competitors or an entire segment of companies. For this reason, only big companies can afford to make such research. Satisfaction surveys remain unavailable for small and medium businesses. Deprived of the ability to assess customer satisfaction, the quality management system becomes defective.

Due to their inherent discreteness, the considered methods do not allow continuous monitoring of the level of customer satisfaction. For this reason, data analysis is limited to one period of time and does not provide insight into trends and dynamics of consumer

Table 1. Methods of customer satisfaction researching recommended by the ISO 10004 standard [19].

Method	Advantages	Constraints
Personal interviews	– contact and personal attention; – the ability to ask more complex and focused questions; – interviewing flexibility; – immediate receipt of information; – the ability to verify the information	– takes more time, slower process; – a more expensive process, especially if the interlocutors are geographically dispersed; – the risk of possible distortion of facts on the part of the interviewer
Telephone interview	– less expensive than face-to-face conversations; – flexibility; – the ability to verify information; – the speed of carrying out; – immediate receipt of information	– impossible to record non-verbal responses (no visual contact); – the risk of distortion of facts on the interviewer part; – receiving information is limited by the short duration of the conversation (20–25 min); – refusal of clients to participate in interviews
Discussion groups	– lower costs compared to personal interviews; – semi-structured questions; – spontaneous.responses caused by a collective approach	– requires an experienced coordinator and appropriate equipment; – the result depends on the familiarity of the participants with this technique; – difficulties arise if clients are scattered over a large area
Correspondence research	– low costs; – coverage of a geographically remote group of respondents is possible; – there is no distortion of facts on the interviewer part; – a high level of unification; – the relative ease of implementation	– the response rate may be insignificant; – an independent choice of respondents can lead to obtaining a distorted sample that does not reflect the studied population; – may be challenging to understand unclear formulated questions; – no control of behavior while filling out the questionnaires; – takes longer to collect data

(continued)

Table 1. (*continued*)

Method	Advantages	Constraints
On-line survey	– low costs; – questions are prepared in advance; – no distortion of facts on the part of the interviewer; – high level of unification/comparability; – fast implementation; – ease of assessment	– low response rate; – lack of control over behavior while filling out the questionnaires; – delays in receiving data; – high probability of interruption of the questionnaire in case of unclearly formulated questions; – assumes that customers have equipment and are familiar with the methodology

sentiment. It also affects the timeliness of making management decisions, which depends on the frequency of receiving up-to-date data on customers' opinions and their emotions. For this reason, the adjustment of processes in the organization will be made with a time lag. On the other hand, frequent surveys and questionnaires can cause clients' negative feelings towards the company. By their nature, the considered methods are more suitable for conducting marketing research when developing a new service.

The selection and application of various scales for assessing customer satisfaction and the subjectivity of respondents' perceptions are still the subjects of the dispute. The resulting customer satisfaction scores are expressed in the form of abstract indices of customer satisfaction, which have no dimension and, in most cases, are calculated as arithmetic mean overall ratings. These indices are difficult to understand, compare and interpret research results.

5 Kano's Model and the Proposed Approach

For a better understanding of the concept of "customer satisfaction", let's consider the Kano model [20], which connects the level of customer satisfaction with the satisfaction of their expectations (Fig. 2). This model divides all aspects of a service into four categories based on their impact on customer satisfaction.

- Requirements for infrastructure characteristics - this category includes service aspects that customers do not attach much importance. Their improvement does not have a significant impact on their satisfaction. However, they are necessary for the operation of the company or the operation of the product. The company may consider reducing the cost of improving them by reallocating resources to improve those with the most remarkable customer value.
- Implied requirements - this category includes the service aspects that the customer expects. They are essential to customers, but once a certain level of improvement has

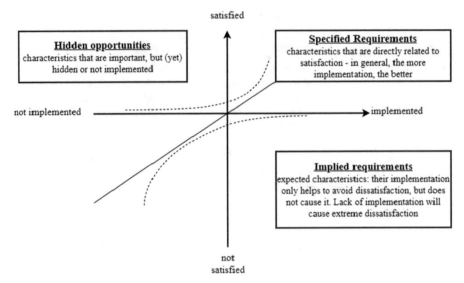

Fig. 2. Model of Kano.

been achieved, further improvement does not affect customer satisfaction any more. The indicators of aspects of this group should not fall below the acceptable level, otherwise this inevitably leads to a decrease in customer satisfaction. The presence of aspects of this category with acceptable indicators prevents customer dissatisfaction. They are not clearly expressed, but they are still very important.

– Specified requirements - this category includes service aspects that directly affect customer satisfaction or dissatisfaction. The better their execution, the higher the customer satisfaction. Customers attach great importance to them, so they require much attention. The quality level of the aspects of this group should be constantly monitored, and in the long term, increased.

– Hidden opportunities - this category includes service aspects that are not recognized or considered necessary at the moment but which can significantly increase the level of satisfaction in general if used correctly. They offer future development opportunities and competitive advantages. The lack of such aspects of the service is not frustrating as they are not yet realized. However, their presence can have a significant impact on satisfaction. It should be noted that such characteristics may eventually become a group of established requirements. And it is to this category that the accounting of emotionally colored information of customers in the management of the process of providing services belongs to.

By categorizing the various aspects of a service as described above, companies can prioritize their improvement efforts. The identification of the service aspects is carried out during the customer satisfaction survey. However, the number of aspects can be very large, so combining them into aspect groups according to their similarity makes sense. When conducting a quantitative study, satisfaction is assessed for individual aspect groups and for products in general.

To assess customer satisfaction with services, it is proposed to modernize the Kano model, emphasizing including into the category of hidden opportunities the consideration of emotionally colored customer information.

Affective computing technologies, which are a field of artificial intelligence aimed at analyzing and processing emotions, can be used to account for customer emotions in the process of providing services. To assess the level of customer satisfaction in the service sector, it is proposed to analyze textual reviews in natural language published on the Internet.

Analysis of data processing methods based on artificial intelligence technologies revealed several difficulties lying in the way of their practical application:

- Imperfection of algorithms for searching customer reviews on the Internet;
- Imperfection of methods for filtering irrelevant messages;
- Imperfection of methods for analyzing text data.

In this regard, an approach to assessing the level of customer satisfaction in the service sector is proposed, based on the developed models and methods for analyzing semi-structured information from natural language texts described by the authors in [21–23].

To develop effective management decisions based on assessing customer satisfaction, it is necessary to consider the following stages: collecting text reviews on the Internet, preprocessing customer reviews, analyzing the sentiment of reviews, analyzing emotionally colored information from customer reviews, analyzing aspects of services from reviews, assessing the level of satisfaction and building decision rules based on customer satisfaction.

6 Implementation

To achieve these goals, we used software developed by the authors in the C# language based on algorithms for classifying sentiment and analyzing emotionally colored information. To train the classifier and assess the accuracy, a sample of reviews in Russian from the top.hotels.ru portal was used, including 304 positive reviews and 850 negative reviews. To assess the effect of lemmatization during text preprocessing on the classification accuracy, the LemmaGen library for C# was used [24]. Cross-validation was used to assess the generalizing ability of the algorithm. A set consisting of 10 partitions of the original sample, each of which, in turn, consisted of two subsamples: training and control, was fixed. For each partition, the algorithm was adjusted according to the training subsample, then its average error was estimated on the objects of the control subsample. The average error assessed the cross-validation for all partitions in the control subsamples. For the bagging (or bootstrap aggregating) algorithm, the admissible error of the classifier is equal to $e = 25\%$. An accuracy metric is used to assess the classification accuracy of each control block.

The results of the computational experiment are presented in Table 2. For the naive Bayes classifier, the best result was obtained by the multinomial model - 86.83%. Bernoulli model reached 86.49%. The bagging algorithm has shown a positive impact on classification accuracy. Bagging allowed to increase the classification accuracy up to 87.69%.

Table 2. Experimental results.

Naive Bayes classifier	
Bernoulli model	86.49%
Multinomial model	86.83%
Multinomial model with synonyms accounting	86.93%
Multinomial model, with excluded words shorter than 3 letters	86.40%
Bagging, Bernoulli model (e = 25%)	86.82%
Bagging, multinomial model (e = 25%)	87.69%
Support-vector machine (SVM)	
Linear kernel, criteria absence/presence of a word	87.69%
Linear kernel, criteria frequency of the word usage	85.00%
Linear kernel, criteria absence/presence of a word, without lemmatization	87.07%
Linear kernel, criteria absence/presence of a word with excluded words shorter than 3 letters	88.21%
Linear kernel, criteria absence/presence of a word with synonyms accounting	87.77%
Polynomial kernel criteria absence/presence of a word	86.73%
Linear kernel, criteria absence/presence of a word with synonyms accounting and excluded words shorter than three letters	88.30%

Support vector machine (SVM) showed the best result on a linear kernel, taking into account the presence/absence of a word in the document - 87.69%. The classification accuracy, taking into account the frequency of word use in the document on a linear kernel, was 85%. The SVM classification accuracy on the polynomial kernel was 86.73%. The results showed that SVM outperforms the Naive Bayes classifier without using the bagging algorithm inaccuracy. Lemmatization positively affected the classification accuracy - 87.69% versus 87.07% for the SVM method. Without lemmatization, one word with different endings will be identified by the algorithm as different words. Lemmatization and other similar techniques, such as stemming (removing endings from words), can improve classification accuracy. To search for synonyms and take them into account, a synonym dictionary containing 5371 meanings was used. The algorithm for finding and processing synonyms is to replace all synonyms with one of the most frequently encountered concepts in reviews. This modification made it possible to slightly increase the accuracy by 0.08% −0.01% for SVM and Naive Bayes classifier. Perhaps the use of a more voluminous dictionary of synonyms will give a greater increase in classification accuracy. Removing words with less than 3 letters from reviews allowed to increase the sentiment classification accuracy for the SVM method by 0.52% and decrease by 0.43% for the Naive Bayes classifier.

The analysis of emotionally colored information from customer reviews for different aspect groups of services allows us to assess quantitatively and qualitatively the level of customer satisfaction and monitor the dynamics of this indicator. Using decision trees, decision rules are formed in managing the process of providing services and taking into account customer satisfaction levels.

7 Conclusion

The analysis of the existing competitive advantages for service companies was carried out. In a highly competitive fight for market share, the company's main task is the fullest satisfaction of the needs of its customers. This can be better achieved by taking into account their individual characteristics and emotionally colored information. The existing research devoted to accounting emotionally colored information in marketing tasks was analyzed.

An analysis was carried out that revealed a number of the deficiencies in the classical methods of customer satisfaction research.

An approach was proposed to assess customer satisfaction by collecting consumer reviews from the Internet and analyzing them using artificial intelligence and affective computing technologies. To assess customer satisfaction based on their feedback, an analysis of the text's lexical sentiment and emotional coloring was used.

A quantitative and qualitative study of customer reviews from the tophotels.ru portal was carried out for a computational experiment. Analysis of the classification accuracy of reviews showed promising results. Machine learning categorizes reviews with 85%–88% accuracy. Lemmatization improves the accuracy of the classification of reviews in Russian. The bagging algorithm has a positive effect on the classification accuracy. Taking synonyms into account can have a positive effect on the accuracy of the classification.

Acknowledgments. The results presented in the article were obtained within the framework of the state assignment of the Russian Federation No. FEUE-2020-0007 and in the implementation of the RFBR grant 19-07-00709.

References

1. Huang, M.-H.: The theory of emotions in marketing. J. Bus. Psychol. **16**, 239–247 (2001)
2. Gaur, S.S., Herjanto, H., Makkar, M.: Review of emotions research in marketing, 2002–2013. J. Retail. Consum. Serv. **21**(6), 917–923 (2014)
3. Laros, F.J.M., Steenkamp, J.-B.: Emotions in consumer behavior: a hierarchical approach. J. Bus. Res. **58**(10), 1437–1445 (2005)
4. Purcarea, I.-M.: Marketing transformation under the pressure of the new technologies and emotions impact on decision making. Holist. Marketing Manag. J. **10**(4), 13–22 (2020)
5. Bigné, J.E., Andreu, L.: Emotions in segmentation: an empirical study. Ann. Tour. Res. **31**(3), 682–696 (2004)
6. Walsha, G., Shiub, E., Hassanc, L.M., Michaelidoud, N., Beattye, S.E.: Emotions, store-environmental cues, store-choice criteria, and marketing outcomes. J. Bus. Res. **64**(7), 737–744 (2011)

7. Bell, H.A.: A contemporary framework for emotions in consumer decision-making: moving beyond traditional models. Int. J. Bus. Soc. Sci. **2**(17), 12–16 (2011)
8. Rostomyan, A.: The impact of emotions in marketing strategy. In: Ternès, A., Towers, I. (eds.) Internationale Trends in der Markenkommunikation, pp. 119–129. Springer, Wiesbaden (2014). https://doi.org/10.1007/978-3-658-01517-6_9
9. Pentus, K., Mehine, T., Kuusik, A.: Emotions in product package design through combining conjoint analysis with psycho physiological measurements. Procedia Consider. Soc. Behav. Sci. **148**(25), 280–290 (2014)
10. Achar, C., So, J., Agrawal, N., Duhachek, A.: What we feel and why we buy: the influence of emotions on consumer decision-making. Curr. Opin. Psychol. **10**, 166–170 (2016)
11. White, C.J.: The impact of emotions on service quality, satisfaction, and positive word-of-mouth intentions over time. J. Mark. Manag. **26**(5–6), 381–394 (2010)
12. Kemp, E.A., Borders, A.L., Anaza, N.A., Johnston, W.J.: The heart in organizational buying: marketers' understanding of emotions and decision-making of buyers. J. Bus. Ind. Mark. **33**(1), 19–28 (2018)
13. Consoli, D.: Emotions that influence purchase decisions and their electronic processing. Annales Universitatis Apulensis Series Oeconomica **11**(2) (2009)
14. Taylor, S.A., Ishida, C., Novak Donovan, L.A.: Considering the role of affect and anticipated emotions in the formation of consumer loyalty intentions **33**(10), 814–829 (2016)
15. Le, D., Pratt, M., Wang, Y., Scott, N., Lohmann, G.: How to win the consumer's heart? Exploring appraisal determinants of consumer pre-consumption emotions. Int. J. Hosp. Manag. **88**, 102542 (2020)
16. Poels, K., Dewitte, S.: The role of emotions in advertising: a call to action. J. Advert. **48**(1), 81–90 (2019)
17. State Standard R ISO 9001-2011: Quality management systems. Requirements, p. 36. Standartinform, Moscow (2012)
18. State Standard R 54732-2011/ISO/TS 10004: 2010. Quality management. Customer Satisfaction. Monitoring and Measurement Guidelines, p. 28. Standartinform, Moscow (2010)
19. Deming, E.: Way out of the Crisis, p. 497. Alba, Tver (1994)
20. Kondo, Y.: Company-Wide Quality Management: trans. from English N. Novgorod: SMTs Priority, p. 236 (2002)
21. Yusupova, N.I., Bogdanova, D.R., Komendantova, N.P.: Artificial intelligence tools for analyzing emotionally colored information from customer reviews in the service sector. IOP Conf. Ser. Mater. Sci. Eng. **1069**, 01 (2013)
22. Kovács, G., Bogdanova, D., Yussupova, N., Boyko, M.: Informatics tools, AI models and methods used for automatic analysis of customer satisfaction. Stud. Inform. Control (SIC) **24**(3), 261–270 (2015)
23. Yussupova, N., Kovács, G., Boyko, M., Bogdanova, D.: Models and methods for quality management based on artificial intelligence applications. J. Appl. Sci. Acta Polytechnica Hungarica **13**(3), 45–60 (2016)
24. Lemmatise. http://lemmatise.ijs.si/Software/Version3. Accessed 21 May 2019

Approaches to the Intellectualization of Decision Support in Irrigated Agriculture Based on Self-organizing Neural Networks

G. Kamyshova[1]([✉]) [ID], S. Ignar[2] [ID], A. Kravchuk[3] [ID], and N. Terekhova[3] [ID]

[1] Financial University, 49/2, Leningradskiy pr-d, Moscow 125167, Russia
gnkamyshova@fa.ru
[2] Warsaw University of Life Sciences, 159, Nowoursynowska Street, 02-776 Warsaw, Poland
[3] Saratov State Agrarian University, Theatralnaia pl., Saratov 410012, Russia

Abstract. In this paper, we consider the results of research on some approaches to decision-making in irrigated agriculture - Intelligent Systems. The traditional resources for increasing the efficiency of agriculture have practically dried up. Achieving a qualitatively new level of agricultural development is possible only with the use of modern digital and intelligent technologies, because of traditional resources for increasing efficiency do not lead to a result. Global climate change and the need for agriculture in risky farming areas necessitate decision support systems for effective governance and sustainable agricultural development. The rapid development of data science and artificial intelligence has led to a new stage in the development of decision support systems - intellectualization. We have proposed to put the Kohonen neural network (Self-organizing map) as the basis for an intelligent decision support system for agro-climatic resources. At the same time, the use of modern software makes it possible to implement complex mathematical models, while new tools for creating intelligent systems such as remote sensing, cloud technologies, big data processing systems and machine learning allow more accurate management decisions at all levels of agricultural production.

Keywords: Intellectualization · Decision support · Irrigated agriculture

1 Introduction

The role of agriculture in the economy of most countries is constantly growing. The reason for this is, on the one hand, an increase in population, and on the other, an increase in the level of urbanization. Wherein, the volume of resources (land and water) required to produce products does not increase. The traditional resources for increasing the efficiency of agriculture (mechanization, development of new lands) have practically dried up. Achieving a qualitatively new level of development of the agro-industrial complex is possible only with the use of modern digital technologies in agricultural management. Digitalization and intellectualization have been made possible in a broad sense by accelerating the development of technology. The fourth technological revolution

A. Gibadullin (Ed.): DITEM 2021, LNNS 432, pp. 159–169, 2022.
https://doi.org/10.1007/978-3-030-97730-6_14

involves a large-scale development and deployment of "smart" systems, equipment, technology, and infrastructure in various fields, particularly agriculture. For example, according to the analysis of the Ministry of Agriculture of Russia, over 50 percent of the costs of agricultural enterprises can be optimized using digital technologies. At the same time, Russia ranks 15th in the world in terms of agricultural digitalization. Only about 10% of Russian farmers use digital technologies, while in the EU countries it is about 80% of farmers, and in the USA - 60% [1].

Management of agrarian systems can be divided into strategic and operational, while they are characterized by insufficient information security, imperfect methods of information processing and decision-making, limited time for making control actions. The development of mathematical methods and computer technology, the emergence of microprocessors and their improvement make it possible to largely remove restrictions on the quality of control. Digital, "smart" agriculture in general and irrigated agriculture in particular, requires the development of new tools that allow making increasingly complex decisions and systematizing data.

Thus, the development of decision support systems is the main task of smart farming. An overview of the applications of artificial intelligence in agriculture is presented in [2, 3] and includes forecasting in agriculture, intelligent monitoring of diseases and pests, and weed control.

Further development of decision support systems in agriculture in general and in irrigated agriculture is their intellectualization. The process of technological development of computer technology, software and related mathematical methods dictates the conditions for the development of appropriate decision support systems. The process of evolution goes from simple information and reference systems to systems based on mathematical simulation models that allow to form solutions. Such decision support systems for irrigated agriculture have been developed by Russian and foreign scientists. The emergence and rapid development of data science and artificial intelligence has led to a new stage in the development of DSS - intellectualization. Work in this direction is already underway [4–6]. In [5], an automated decision support system for irrigation control in a specific field of crops is proposed, based on both climatic and soil variables provided by meteorological stations and soil sensors. For example, FieldNET Decision Support System provides a fast and convenient solution for water, fertilizer, and chemical applications with remote control. The IrrigaSy system, developed in [7], is a decision support system (DSS) for irrigation water management based on open source online tools. The LCIS DSS, developed in [8], an irrigation supporting system for water use efficiency improvement in precision agriculture One approach is to integrate irrigation decision support systems [9] with automated irrigation systems.

However, the existing solutions do not fully cover the needs of the agricultural sector, in particular, irrigated agriculture. Namely, agro-climatic resources are one of the key factors for the efficiency of irrigated agriculture. In countries with large territories and diversification in agro-climatic zones, their correct assessment is critically important. The purpose of this work is to develop neural network approaches to constructing an intelligent decision support system for irrigated agriculture.

2   Materials and Methods

Most of the traditional Russian agricultural crops, such as soybeans and corn, require irrigation, while favorable regions can have an arid climate. So, for example, irrigation norms for soybeans can vary within the range: 1300–4900 m3/ha according to data only within one region, such as the Saratov region. And this trend, in the light of global climate change, can only get worse. Some analysis of forecasts of agroclimatic changes according to the most severe scenario for the Middle and Lower Volga regions is shown in Table 1 [10].

Table 1. Forecast of agro climatic changes according to the most severe scenario of global warming for the Middle and Lower Volga regions.

Region	Sum of active temperatures, °C			The amount of precipitation, mm		
	2000 years	2046–2065 years	2081–2100 years	2000 years	2046–2065 years	2081–2100 years
Samara and Saratov right bank regions	2855	3281	3643	315	220	159
Saratov left bank region	3106	3532	3894	217	178	117
Volgograd region	3395	3831	4193	207	127	66
Astrakhan region	3837	4263	4625	129	55	10

Traditionally, in decision support systems for agro climatic parameters, classical statistical methods of correlation and analysis of variance are used, while data mining models and artificial intelligence methods are used in intelligent decision support systems. Artificial intelligence tools - fuzzy logic, decision trees, artificial neural networks, etc. can be used as the basis for such systems. We will use neural network algorithms as the basis for building intelligent decision support systems.

An artificial neural network, as the most famous model of artificial intelligence, is a set of neurons with a specific architecture formed based on the relationships between neurons in different layers. The main element of a neural network is a neuron.

The mathematical formalization [11] of an artificial neuron is the adder equation:

$$s_k = \sum_{j=1}^{n} w_{j,k} \cdot x_j + b_k \tag{1}$$

and activation block equation

$$y_k = \phi(s_k), \tag{2}$$

Where, for k-th neuron, $x_1, x_2, ..., x_n$ - input signals; $w_{1,k}, w_{2,k}, ..., w_{n,k}$ - synoptic weights; b_k - reference signal level; s_k - linear adder output; $\phi(s_k)$ - activation block conversion function; y_k - output signal. Figure 1 shows a functional diagram of an artificial neuron model and a multilayer neural network.

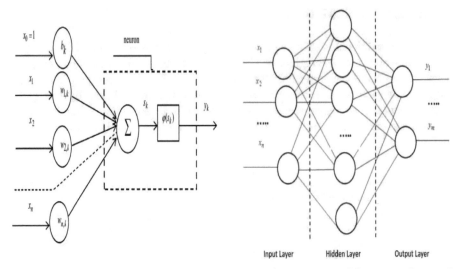

Fig. 1. Functional diagrams of an artificial neuron model (left) and a multilayer neural network (right).

A set of neurons connected by inputs and outputs makes up an artificial neural network. These networks can have different architectures, which leads to different types of ANNs. Different types of ANNs are used to solve different problems. We will consider the so-called self-organizing neural networks (neural network architecture of unsupervised learning), namely self-organizing Kohonen maps. They allow identifying clusters of input vectors that have some common properties. The main tasks solved with their help are best suited for the formation of intelligent DSS, namely, we will consider two of them:

– Forecasting agro-climatic resources by correlating new objects to one of the clusters, it is possible to predict the behavior of objects, since it will be similar to the behavior of cluster objects;
– Detection of anomalies in agro-climatic resources by analyzing clusters with a small number of objects helps to identify anomalies.

2.1 Mathematical Model

We use a feed-forward network architecture. The structure of a neural network has two layers - an input and a single layer of neurons (Kohonen layer) without displacement coefficients (Fig. 2) [12].

Traditionally, for networks of this architecture, three main stages are considered: training, cluster analysis and practical use. The general scheme and mathematical implementation of the Kohonen network learning algorithm is well developed and can be summarized in the following steps:

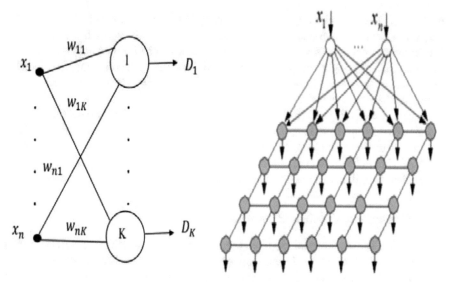

Fig. 2. Kohonen's neural network: network connections (right), general view (left).

- Initialization process. Initially, the original synaptic weight vectors are filled with random values;
- The process of selecting a training signal. Reading a set of input values from the database and choosing, with a certain probability, an excitation vector to be applied to the lattice of neurons;
- The process of finding the maximum similarity. Search for the "optimal" neuron according to the criterion of the minimum Euclidean distance;
- Correction process. The weights of the neurons located in the vicinity of the "optimal" neuron are being adjusted;
- Go to step 2.

The steps of the algorithm are repeated until the output values of the network are stabilized with the specified accuracy.

At the training stage, the distances d_j from the training samples to each neuron are calculated using the formula [12]:

$$d_j = \sum_{i=1}^{n} (w_{ij} - x_i)^2 \qquad (3)$$

w_{ij} – the weighting factor connecting the input vector x_i to the j-th cluster element.

Then, over all distances, the minimum

$$d = min_{1 \leq j \leq n}(d_j) \qquad (4)$$

is calculated and the weight coefficients of the corresponding neurons are adjusted according to one of the standard algorithms, for example,

$$w_{ij}(n + 1) = w_{ij}(n) + s(n)(x_i - w_{ij}(n)) \qquad (5)$$

Where $w_{ij}(n)$ – the weighting factor connecting the input vector x_i to the j–th cluster element at iteration n, and $s(n)$ – a learning rate factor with values from zero to one. The learning process ends when a certain criterion is reached, for example, the value of the change in the weight coefficients at the next iteration - if it is less than the specified value, then the process is completed. This is not the only algorithm that can be applied, there are other ways as well.

A cluster analysis procedure is applied to the trained neural network. The Kohonen neural network application algorithm consists either in correlation to one of the existing clusters, or in the conclusion that such a correlation is impossible for a new example. Moreover, in the first case, the description of the cluster and the corresponding solutions are applicable to the newly analyzed example.

2.2 Materials and Problem Statement

As databases for model validation, we use data from official climate reference books, meteorological observations in the Roshydromet system, data from field experiments conducted on the basis of irrigation systems in the Saratov region of Russia, materials from the work of Russian scientists (for example, [13, 14]. The main climatic parameters of the irrigated zone are presented in Table 2.

Table 2. The main parameters of the climate of irrigated lands in the Saratov region.

No	Parameter	Value
1	Winter temperatures, °C	−10...−15
2	Summer temperatures, °C	21.6...22.6
3	The sum of effective temperatures, °C	2400...3100
4	Frost-free period, days	130...170
5	The period of active vegetation of crops, days	165...180
6	The depth of soil freezing, sm	100...145
7	Average annual rainfall, mm	310...500
8	Spring moisture reserves in a meter layer of soil, m^3/h	600...1700

The sums of air humidity deficits in the warm season increase from north to south from 1800 to 2200 mm. On the contrary, the sums of active temperatures decrease in the same direction from 3500 to 3000 °C. The amount of precipitation in the warm season for the Saratov Trans-Volga region in the northern direction increases from 150 mm in the extreme southeast of the region to 250 mm, and in the northeast region - up to 275 mm.

The most important environmental problem facing humanity in the 21st century is the rise in temperature of the climate system of our planet - the so-called global warming. Over the past thirty years on the territory of the Saratov region during the main growing season of grain crops (May-July), the climate aridity has increased: the average air temperature has increased by 0.80 °C, while the amount of precipitation has decreased by 10 mm.

The amount of autumn-winter precipitation (November-February) increased by 20% and the amount of precipitation decreased in August. Thus, the analysis shows that even within one region, the differentiation of agro climatic resources is high. All this urgently requires the development of intelligent decision support systems that allow predicting agro-climatic resources and detecting anomalies for optimal planning of irrigated crops and material resources.

3   Results and Discussions

The main agro climatic parameters, such as the sum of active temperatures, the sum of precipitation, the moisture coefficient, etc. determine the assessment of biological productivity on what basis the yield of agricultural crops is predicted, which we will take as the basis for clustering.

We are considering three clusters by categories: low, medium and high biological productivity. The number of clusters can be larger. The work of an intelligent algorithm in the structure of an intelligent system is as follows. We will supply data for each observation polygon at the input of the network (Table 3).

Table 3. Data structure.

Polygon number	Sum of active temperatures X_1	Average annual rainfall X_2	Moisture factor X_3	...	Biological productivity assessment X_n
i	a_{i1}	a_{i2}	a_{i3}	...	a_{in}

Here a_{ij} is a character of the j-th parameter on the i-th polygon.

Task intelligent algorithm would consist in agro-climatic data clustering parameters by three categories of biological evaluation of productivity, enabling to consider the characteristic properties. And, to assess the possible yield for the main irrigated crops applicable for practical use - low, medium and high.

Consider the main agro climatic parameters, such as the sum of active temperatures, the amount of precipitation, the sum of the average daily moisture deficit values, and the moisture coefficient (Table 4).

Let's simulate a self-organizing Kohonen map in Matlab [15]. The network architecture and learning process are shown in Fig. 3a.

We obtain the following characteristics of the object of study: the location of the neurons in the topology and the amount of training data associated with each of the neurons - cluster centers (Hits) Fig. 3c, the distance between neighboring neurons (SOM Neighbor Weight Distances) Fig. 3b and the neural network weight coefficients Fig. 3d. Analysis of the trained network based on regional data shows clusters of biological productivity

Table 4. Data example.

Polygon number	Sum of active temperatures	Average annual rainfall	Sum of the average daily moisture deficit values	Moisture factor	Biological productivity assessment
1	2684	594	1730	0.34	High
2	2959	347	2280	0.15	Low
3	3430	290	2785	0.1	Low
4	2693	654	1726	0.38	Medium
5	3036	498	1716	0.29	High
6	3498	353	2430	0.15	Low
7	2518	583	1310	0.45	High
8	2888	616	2042	0.3	Medium
9	2530	529	1561	0.34	Medium
10	2996	336	2206	0.15	Low
11	2671	474	1684	0.28	Medium

assessment: low 18%, average 49% and high 33%. The trained neural network is ready for implementation into a decision support system.

The trained neural network is tested using a specific year dataset with specific parameters Sum of active temperatures, Average annual rainfall and etc., determining the optimal biological productivity. The quality assessment of the constructed network can be assessed using the so-called ROC-analysis method (Receiver Operator Characteristic).

The basis of this analysis is the construction of the so-called ROC-curve. It shows the dependence of the number of correctly classified positive examples on the number of incorrectly classified negative examples. ROC-analysis data with a confidence interval of 95% are shown in Table 5.

Area under the ROC curve - 0.980; the standard error is 0.021; 95% confidence interval 0.895–0.997; significance level - 0.0001.

The constructed neural network model shows good quality and can be included in the control system.

The proposed architecture of an intelligent DSS includes three modules (Fig. 4). The first central module is the databases and knowledge databases that provide the use of computational means of the other two modules of an integral and independent system of knowledge about the environment. The executive system is a module that combines a set of tools that ensure the execution of the generated program. Namely, the application of cluster analysis procedures, the practical use of a neural network and the formation of conclusions based on the knowledge bases of the main irrigated crops. The third module is an intelligent interface that provides communication and system recommendations for the end user.

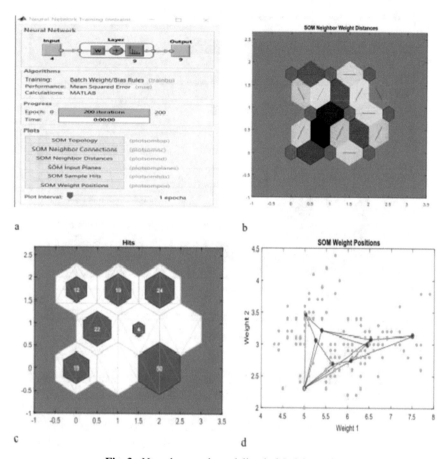

Fig. 3. Neural network modeling in Matlab results.

Table 5. ROC analysis.

Criterion	Sensitivity	95% confidence interval
>0	100	85.1–100
>0	100	85.1–100
>0.003	100	85.1–100
>0.8	96.35	78.2–99.3
>0.85	90.81	70.5–97.3
>0.9	85.38	66.3–95.6
>1	0	0–16.2

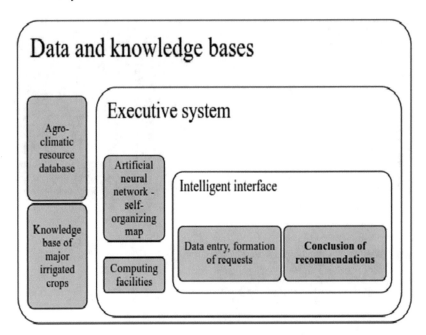

Fig. 4. Architecture of an intelligent DSS.

Forecasting agroclimatic resources in a changing climate is more relevant than ever. Currently, many researchers are moving from physical models to models based on artificial neural networks, for example [16]. Another problematic is the further use of these models in decision support systems. For example, an approach based on the use of Kohonen's neural network and GIS technologies was applied in [17] to cluster the spatial-temporal variability of irrigated areas.

4   Conclusions

Due to global climate changes, zonal differences and instability of weather conditions within one region and the influence of these factors on the size and quality of the yield of irrigated crops, it becomes important to develop decision support tools at different levels of management. In this regard, the central direction in the development of scientific and technological progress in agriculture is to improve the management of irrigated agriculture, increase yields on irrigated lands and resource conservation through the development of intelligent decision support systems, digital information technologies and mathematical models.

Approaches based on the use of neural network technologies lead to the possible implementation of better and more efficient control, and the availability of modern software allows solving complex problems and taking into account dynamic components. The proposed approach to building a DSS based on an intellectual component will allow obtaining more accurate management decisions due to the fact that the ANN-model at the heart of the decision support system allows us to reveal deeper connections between the parameters under study. All this will contribute to increasing the efficiency of irrigated agriculture.

References

1. Trufliak, E., Kurchenko, N., Daibova, L., Kreimer, A., Podushin, Y., Belaia, E.: Monitoring and Forecasting the Scientific and Technological Development of the Agro-Industrial Complex in the Field of Precision Agriculture, Automation and Robotization. KubGAU, Krasnodar (2017)
2. Kujawa, S., Niedbała, G.: Artificial neural networks in agriculture. Agriculture **11**, 497 (2021)
3. Kirtan, J., Aalap, D., Poojan, P., Manan, S.: A comprehensive review on automation in agriculture using artificial intelligence. AI in Agric. **2**, 1–12 (2019)
4. Yakushev, V.P., Yakushev, V.V., Matvienko, D.A.: Intelligent support systems for technological solutions in precision farming. Zemledelie **1**, 33–37 (2020)
5. Navarro-Hellin, H., Martinez-del-Ricon, J., Domingo-Miguel, R., Soto-Valles, F., Torres-Sances, R.: A decision support Haykin S 2016. In: Handbook of Neural Networks, 2 edn. 1104 System for Managing Irrigation in Agriculture. Computers and Electronics in Agriculture, vol. 124, pp. 121–131 (2016)
6. Giusti, E., Marsili-Libelli, S.: A fuzzy decision support system for irrigation and water conservation in agriculture. Environ. Model. Softw. **63**, 73–86 (2014)
7. Simionesei, L., Ramos, T., Palma, J., Oliveira, A., Neves, R.: IrrigaSys: a web-based irrigation decision support system based on open source data and technology. Comput. Electron. Agric. **178**, 105822 (2020)
8. Bonfante, E., et al.: CIS DSS—an irrigation supporting system for water use efficiency improvement in precision agriculture: a maize case study. Agric. Syst. **176**, 102646 (2019)
9. Wang, E., et al.: Development of a closed-loop irrigation system for sugarcane farms using the internet of things. Comput. Electron. Agric. **172**, 105376 (2020)
10. Korsak, V.V., Kravchuk, A.V., Prokopec, R.V., Nikishanov, A.N., Arzhanuhina, E.V.: Scenarios of global warming and forecasts of changes in agroclimatic resources of the Volga region. Agrarny nauchny zhurnal **1**, 51–55 (2108)
11. Haykin, S.: Handbook of Neural networks, 2 edn., p. 1104 (2016)
12. Kohonen, T.: Self-organizing maps, vol. 655 (2008)
13. Roshydromet. Federal Service for Hydrometeorology and Environmental Monitoring, Electronic resource. http://www.meteorf.ru/. Accessed 15 Oct 2021
14. Ivanova, G.F., Levitckaia, N.G., Orlova, I.A.: Assessment of the current state of agroclimatic resources of the Saratov region. Izvestia Saratovskogo universiteta. Novaya seriya. Nauki o Zemle **13**(2), 10–13 (2013)
15. Beale, M., Hagan, M., Demuth, H.: Neural Network ToolboxTM User's Guide, vol. 410. The Math Works Inc., Natick (2010)
16. Ise, T., Oba, Y.: forecasting climatic trends using neural networks: an experimental study using global historical data. Front. Robot. AI. **6**, 32 (2019)
17. Kamyshova, G.N., Soloviov, D.A., Kolganov, D.A., Korsak, V.V., Terekhova, N.N.: Neuro-modeling in irrigation management for sustainable agriculture. Adv. Dyn. Syst. Appl. **16**(1), 159–170 (2021)

Modeling of Process of Radioactive Graphite Processing in Gas-Generating Installation

Nikolay Barbin, Anton Kobelev[✉], Vladimir Lugovkin, Dmitrij Terent'ev,
and Stanislav Titiov

Ural Institute of the State Fire Service EMERCOM of Russia, Ekaterinburg 620062, Russia
antonkobelev85@mail.ru

Abstract. Using the "GRAFIT-GAS" software developed in the CODESYS integrated environment, modeling of the process of reactor graphite processing in a gas generator was carried out. The composition of the generator gas used in the model was determined by the method of thermodynamic modeling of the radioactive graphite – water vapor system using the TERRA software package. The scheme of the gas generator installation and its description are presented. The gasification unit is based on a gas generation scheme with a direct gasification scheme for raw materials based on electric heating. The interface of the "GRAFIT-GAS" program is presented (start window, window for setting the temperature of the gas generator furnace, a graph of the temperature in the gas generator furnace, a graph of the gas composition at the outlet of the furnace). The main chemical reactions occurring in a gas generator are considered. Five stages that simulate the operation of a gas generator are defined.

Keywords: Radioactive graphite · Gas generation · Water vapor

1 Introduction

The nuclear power possesses almost unlimited fuel resource compared to traditional hydrocarbon power.

At present three fundamental problems defining the relation of society to development of nuclear power as potentially dangerous technology exist: risk of severe accidents, treatment of radioactive waste (including spent nuclear fuel), non-proliferation of the sharing materials (risk of global nuclear terrorism).

Spent nuclear fuel compared to radioactive waste after processing can serve as new nuclear fuel for the NPP. The characteristic of radioactive wastes is that the only acceptable way of their relative neutralization is a long-term storage for disintegration of the radionuclides which are contained in them for a long time.

Among all mass of the stored radioactive waste graphite has a specific place. After long radiation, graphite does not gain any properties which could create the field of its useful application [1–4].

In the nuclear industry, graphite is used in the form of non-replaceable products (in the form of graphite blocks) and replaceable elements: contact rings between masonry and technological channels and etc. [1–4].

A. Gibadullin (Ed.): DITEM 2021, LNNS 432, pp. 170–180, 2022.
https://doi.org/10.1007/978-3-030-97730-6_15

In Russia 13 industrial uranium-graphite reactors and 20 power uranium-graphite reactors were constructed: reactor at Obninsk NPP, 2 AMB-100 and AMB-200 reactors at Beloyarsk NPP, 4 EGP reactors at Bilibino NPP, 11 RBMK reactors at Leningrad, Kursk and Smolensk NPP. At the present moment industrial uranium-graphite reactors of Beloyarsk, Leningrad, Biblinsky, Obninsk NPP are stopped and works on decommissioning are conducted. The service life of the RBMK and EGP power reactors comes to the end. For the next period of about 10–15 years, the resource of most blocks will be exhausted taking into account the extension of their service life (Fig. 1) [5].

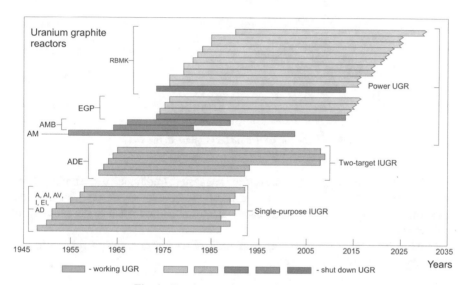

Fig. 1. Russian uranium-graphite reactors.

The total amount of the irradiated reactor graphite in Russia is about 60 thousand tons. Besides Russia the problem of treatment of the irradiated reactor graphite is relevant for Great Britain – more than 86 thousand tons, the USA – more than 55 thousand tons and France – more than 23 thousand tons. The total amount of the irradiated graphite which is stored around the world numbers nearly 250 thousand tons (Fig. 2) [6].

Treatment of the irradiated graphite, including its conditions for disposal is one of critical tasks [5].

The solution of a question on disposal of reactor graphite, is aggravated by the existence of long-living radionuclides in its composition (for example, a half-life period 243Am–7370 years) and also the fact that graphite is a flammable material. This fact is aggravated by the existence of the stored Vigner's energy in the irradiated graphite [1].

In the IAEA document it is noted that concerted strategy for graphite processing is not accepted in the world, but the majority of the countries intend rather to dispose graphite in geological formations, than to deactivate it. To some extent such decision is caused by very slow progress in creation of effective processing technology [5].

Nowadays the most perspective ways of treatment of the fulfilled graphite materials are burning [1].

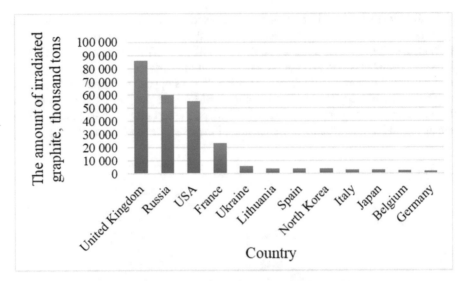

Fig. 2. Amount of irradiated graphite in the world.

Different ways of graphite combustion are suggested: traditional; in the boiling layer; by means of the plasmochemical reactor, gasification of graphite by means of superheated water vapor (pyrolysis), in fusion of carbonates of alkaline metals in the presence of oxidizer [1].

In the present work the modeling of process of reactor graphite processing in gas-generating installation is carried out.

The purpose of gas-generating processing of reactor graphite is to produce artificial combustible gases as a result of incomplete combustion of carbon-containing substances. Gas-generating process is carried out at temperature not below than 500 °C.

Depending on the method of gas generation (the type of gas supplied to the gas generator unit), the artificial generated gas is divided into air, steam, water and coke.

Steam-air generator gas contains a large amount of carbon monoxide.

Graphite stacks of uranium-graphite reactors of nuclear power plants can be a specific source of raw materials for gas-generating installation [7].

One power unit of the RBMK nuclear power plant contains 1850 tons of reactor graphite. After disintegrating electrochemical processing of RBMK-1000 graphite blocks (removal of the most radioactive outer layer of graphite blocks ~1 mm), approximately $1.5 \cdot 10^6$ m^3 of generator gas can be produced. For the processing of graphite masonry of one RBMK reactor, it will take from 3 to 6 years of continuous operation at gas-generating installation capacity from 20 to 50 m^3/h [7].

The dilution coefficient of residual specific radioactivity upon transition of radioactive material from solid state into gaseous one is proportional to the relation of the corresponding density taking into account mass fraction of radioactive element in gas molecule (1):

$$K = \rho_C \cdot m_{CO} / \rho_{CO} \cdot m_C \qquad (1)$$

where ρ_C – density of reactor graphite; ρ_{CO} – carbon oxide density; m_C – relative atomic mass of carbon; m_{CO} – relative molecular mass of carbon monoxide [7].

Considering that average density of reactor graphite grades equals to 1.7 g/cm^3, and density of carbon monoxide is under normal conditions equals to 1.25 g/l, the coefficient of dilution of residual specific radioactivity will be equal to 3170. Thus, the residual specific radioactivity of generating gas will be 3170 times less than the radioactivity of reactor graphite. Ashes content at the same time will not exceed $0.2 \cdot 10^{-3}\%$, it means that combustible gas will be an eco-friendly product. Permissible concentration of carbon-14 in 1 g of reactor graphite is $7 \cdot 10^{-3}$ Ki/l. This concentration is safe for the environment [7].

2   Materials and Methods

The composition of the generator gas for the steam gas generation process can be determined by thermodynamic modeling of the radioactive graphite - water vapor system using the TERRA software package. The program is designed to determine the composition of phases, thermodynamic and transport properties of arbitrary systems with chemical and phase transformations. For calculations of equilibrium structure of phases and parameters of balance database of properties of individual substances (HSC, IVTANTERMO etc.) was used [8–15]. The results of calculating the fraction of dry gases generated for the system of radioactive graphite - water vapor, in the temperature range from 100 to 600 °C are presented in Table 1.

In Fig. 3 the offered scheme of steam gas-generating processing of radioactive graphite is shown.

Table 1. Distribution of a fraction of the generated dry gases in the temperature range from 100 to 600 °C.

T, °C	CO	H$_2$	CO$_2$	CH$_4$
100	$4.83959 \cdot 10^{-8}$	0.001162903	0.499701887	0.499135162
200	$1.48996 \cdot 10^{-5}$	0.013890033	0.49651126	0.489583807
300	0.00061785	0.071531292	0.481650146	0.446200712
400	0.008002062	0.207917367	0.442017981	0.34206259
500	0.048450384	0.38356176	0.36777161	0.200216246
600	0.157033658	0.512241374	0.254163885	0.076561082

At the first stage, fine radioactive graphite is loaded into the gas generator furnace. The gas generator furnace is heated to temperature of ~600 °C. At the second stage, water vapor is introduced into the gas generator furnace, while water vapor combines with graphite. At the third stage gases (CO$_2$, CO, CH$_4$, H$_2$) are removed from the furnace chamber. In the fourth stage, the generator gas is in the filter system.

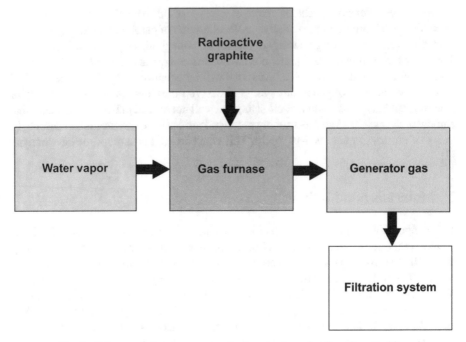

Fig. 3. Scheme of steam gas-generating processing of radioactive graphite.

The gasification unit for the developed of the mathematical model of the process of radioactive graphite processing is based on the scheme of a gas generator with direct gasification of raw materials based on an electric heating device.

3   Results and Discussion

In the mathematical model of the process, the following reactions (2–4) in the gas generator are taken into account:

$$C + H_2O = CO + H_2 \tag{2}$$

$$C + 2H_2O = CO_2 + 2H_2 \tag{3}$$

$$C + 2H_2 = CH_4 \tag{4}$$

The calculation algorithm defines eight stages, five of which simulate the operation of a gas generator:

1. The value of mass of reactor graphite ΔGg reacting in one second (the base case ($\Delta Gg = 0.019554$ kg/s) is set [7]).

In the computer program the dependence (ΔG_g) on a consumption of steam (G_s) and mass of the remained graphite is set (G_g) (5):

$$\Delta Gg = 0.000332 \cdot Gg \cdot (GS)^{0.5} \tag{5}$$

Checking this dependence for the base case with $G_g = 250$ kg and $G_s = 0.058$ kg/s ΔG_g gives 0.02 kg/s.

2. The fractions of graphite mass in each reaction are calculated: α_1, α_2, α_3 ($\alpha_1 + \alpha_2 + \alpha_3 = 1$), considering the generator gas composition (on dry gas at temperature of 600 °C) (6–8):

$$\alpha_1 = 2 \cdot \varphi CO/(1 - \varphi CO_2 + 3 \cdot \varphi CH_4) \tag{6}$$

$$\alpha_2 = 2 \cdot \varphi CO/(1 - \varphi CO_2 + 3 \cdot \varphi CH_4) \tag{7}$$

$$\alpha_3 = 2 \cdot \varphi CO/(1 - \varphi CO_2 + 3 \cdot \varphi CH_4) \tag{8}$$

where φCO, φCO_2, φCH_4 – is the fraction of the corresponding component in the dry generator gas.

The dependence of gas composition on temperature in the range of 500 ... 600 °C is considered (9–12).

$$\varphi CO = 0.001081 \cdot t - 0.4915 \tag{9}$$

$$\varphi CO_2 = -0.00115 \cdot t + 0.9442 \tag{10}$$

$$\varphi CH_4 = -0.00122 \cdot t + 0.8086 \tag{11}$$

$$\varphi H_2 = 0.00128 \cdot t - 0.25636 \tag{12}$$

The values of coefficients are chosen from the data of Table 1 for the range of 500 – 600 °C.

After substitution, expressions for α_i, programmed for calculation on the computer are received (13–15).

$$\alpha_1 = 0.0031 \cdot t + 1.538 \tag{13}$$

$$\alpha_2 = -0.001 \cdot t + 1.121 \tag{14}$$

$$\alpha_3 = -0.0021 \cdot t + 1.427 \tag{15}$$

3. Release of gas components is defined Vi, m³/s (16–19):

$$V_{CO} = 22.4/12 \cdot \Delta G_g \cdot \alpha_1 \tag{16}$$

$$V_{CO_2} = 22.4/12 \cdot \Delta G_g \cdot \alpha_2 \tag{17}$$

$$V_{CH_4} = 22.4/12 \cdot \Delta G_g \cdot \alpha_3 \tag{18}$$

$$V_{H_2} = 22.4/12 \cdot \Delta G_g \cdot (\alpha_1 + 2 \cdot \alpha_2 - 2 \cdot \alpha_3) \tag{19}$$

and theoretical consumption of steam GH_2O (20), kg/s for the reaction (2–4):

$$G_{H_2O} = 18/12 \cdot \Delta G_g \cdot (\alpha_1 + 2 \cdot \alpha_2) \tag{20}$$

4. The remained graphite mass Gg – ΔGg, and an excess consumption of steam are calculated (21):

$$\Delta G_s = G_s - GH_2O \tag{21}$$

5. The composition of wet gas is defined (at the exit from a gas generator) (22):
- total exit of wet gas, m^3/s:

$$V_{w.g.} = V_{CO} + V_{CO_2} + V_{CH_4} + V_{H_2} + \Delta G_s \cdot 22.4/18 \tag{22}$$

content of each component, $V_{CO}/V_{w.g.}$ and etc.
Stages 1..5 model the processes occurring in the control object.
6. The composition and output of dry gas is determined (the control system will cxreceive these data from the generator gas flow meter and gas analyzer) (23):
- total output of dry gas, m^3/s:

$$V_{d.g.} = V_{CO} + V_{CO_2} + V_{CH_4} + V_{H_2} \tag{23}$$

- content of each component, $V_{CO}/V_{d.g.}$ and etc.
7. The mass of carbon in a dry generator gas is determined (the amount of graphite processed in one second, kg/s) (24):

$$\Delta G_g = V_{d.g.} \cdot (\varphi_{CO} + \varphi_{CO_2} + \varphi_{CH_4})/22.4 \cdot 12 \tag{24}$$

8. The mass of the remaining graphite is determined Gg – ΔGg.
The virtual model of the process control system solves the following tasks: regulation of the temperature in the furnace, regulation of the gas flow, calculation the amount of reacted graphite and determining the end of purging. As the present information, data of the temperature in the furnace, steam flow rate, initial mass of loaded graphite, flow rate and composition of dry generator gas are used. Temperature control is carried out by changing the power of the heaters. An electric valve is used to control the flow of water vapor supplied from the steam generator. Direct process control is carried out by a programmable logic controller.
To simulate the operation of the gas generation process control system, a computer program "Model of the processing of radioactive graphite in a gas generator furnace (GRAFIT - GAS)" was developed.
The software "Graphite - Gas" is developed in an integrated environment of CODESYS development. CODESYS (Controller Development System) is a instrumental program complex of industrial automation.
In Fig. 4 the starting window of the program is represented. Four buttons are located on screen: setting up an automatic temperature control system, methodical instructions, reference, exit.

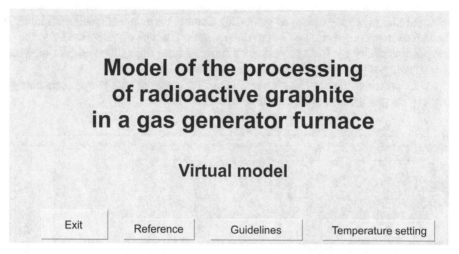

Fig. 4. The starting window of the program.

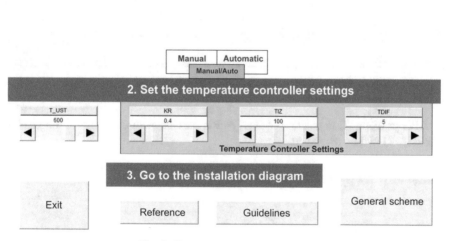

Fig. 5. Temperature setting window.

Figure 5 shows the window - setting up an automatic temperature control system. After pressing the "START" button, the control mode (manual or automatic) is selected and the controller settings are set.

After pressing the General Scheme button on screen the scheme of gas-generating installation will be displayed. Time indicator (s), temperature indicator in the gas-generating furnace (°C), the indicator of the remained mass of graphite (kg), steam flow indicator (kg/s) and steam start button, the indicator of power of heating (%), the button of start of heating of the gas-generating furnace, the button of loading of graphite, the indicator of composition of wet gas, the switch of a type of gases, the indicator of composition of dry gas (m^3/s), condensate flow indicator (kg/s), the indicator of speed of processing of graphite (kg/s) are located on the scheme.

After pressing "Heating" button and then "Temperature" the temperature change graph in the gas-generating furnace (Fig. 6) will be observed.

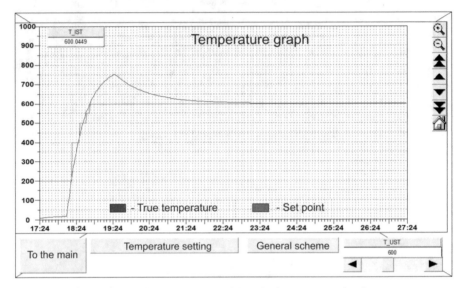

Fig. 6. The fraph of temperature change in the gas-generating furnace.

After pressing "Composition of Gas" button the change of gas composition graph in time (Fig. 7) will be displayed.

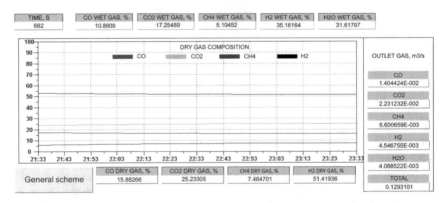

Fig. 7. The graph of change of gas composition in the gas-generating furnace.

4 Conclusion

Thus, the work simulated the process of reactor graphite processing in a gas generator. The scheme of steam gas generator processing of radioactive graphite is described. The description of the computer program "Model of Process of Processing of Radioactive Graphite in the Gas-generating Furnace" and its model is given. The program allows to simulate the process of radioactive graphite processing by using the Gas-generator study scheme, the trend of changes in the generated gas mixture composition and visualizations demonstrating the work of the basic units of the control system. The virtual model of the gas generator set allows to calculate the current composition of the generator gas and its consumption, the amount of processed graphite, taking into account the effect of temperature on the gas mixture composition. It allows to set the temperature controller settings, the load mass of reactor graphite, the flow rate of steam supplied to the gas generator installation. The calculation results are reflected in the form of a change trend of the main process parameters and tables over time.

References

1. Tsyganov, A.A., et al.: The problems of utilizing graphite of stopped graphite-uranium reactors. Bull. Tomsk Polytechn. Univ. **310**(2), 94–98 (2007)
2. Bulanenko, V.I., Frolov, V.V., Nikolayev, A.G.: Radiation characteristics of graphite in carbonuranium reactors removed from service. Atomnaya Energiya **81**(4), 304–306 (1996)
3. Pavluk, A.O., et al.: Radiometry of radiation fields in graphite stacks of stopped commercial carbon uranium reactors. Bull. Tomsk Polytechn. Univ. **309**(3), 68–72 (2006)
4. Boyko, V.I., Shidlovki, V.V., Gavrilov, P.M., Nesterov, V.N., Shamanin, I.V., Ratman, A.V.: Estimation reactor graphite resource in control and protection system cell and in terms of thermophysical properties degradation. Izvestiya Vuzov. Ser. Yadernaya Energetika **3**, 94–103 (2005)
5. Conditioning of Reactor Graphite from Decommissioned Uranium–Graphite Reactors for Disposal. http://www.atomic-energy.ru/articles/2016/06/08/66585. Accessed 04 Oct 2021
6. Dorofeev, A.N., et al.: On reactor graphite disposal. Radioact. Waste **2**(7), 18–30 (2019)
7. Skachek., M.A.: Radioactive Components of Nuclear Power Plants: Management, Processing, and Localization: Textbook for Higher Schools. Moskow Energineering Institute, Moscow (2014)
8. Belov, G.V., Iorish, V.S., Yungman, V.S.: Ivtanthermo for windows - database on thermodynamic properties and related software. Calphad **23**(2), 173–180 (1999)
9. Belov, G.V., Iorish, V.S., Yungman, V.S.: Simulation of equilibrium states of thermodynamic systems using ivtantermo for windows. High Temp. **38**(2), 191–196 (2000)
10. Belov, G.V.: Thermodynamic analysis of combustion products at high temperature and pressure. Propellants Explos. Pyrotech. **23**(2), 86–89 (1998)
11. Belov, G.V.: Influence of thermodynamic and thermochemical data error on calculated equilibrium composition. Berichte der bunsengesellschaft/Phys. Chem. Chem. Phys. **102**(12), 1874–1879 (1998)
12. Aristova, N.M., Belov, G.V., Morozov, I.V., Sineva, M.A.: Thermodynamic properties of condensed uranium dioxide. High Temp. **56**(5), 652–661 (2018)
13. Belov, G.V.: Determining the phase composition of complex thermodynamic systems. Russ. J. Phys. Chem. A **93**(6), 1017–1023 (2019)

14. Belov, G.V., et al.: The ivtanthermo-online database for thermodynamic properties of individual substances with web interface. J. Phys. Conf. Ser. **946**, 012120 (2018)
15. Barbin, N.M., Kobelev, A.M., Terent'ev, D.I., Alekseev, S.G.: Thermodynamic analysis of the oxidation of radioactive graphite in a multicomponent melt in an inert atmosphere. Russ. Metallur. (Metally) **2018**(8), 700–706 (2018). https://doi.org/10.1134/S0036029518080025

Bohr's Complementarity Principle and Management Decision Making

I. S. Klimenko[✉], E. A. Palkin, and L. V. Sharapova

Russian New University, Moscow, Russia
igor.k41@yandex.ru, palkin@rosnou.ru

Abstract. The article deals with the problem of choosing management decisions in conditions of lack of information, time and resources. The heterogeneous relationship of the parameters characterizing the degree of uncertainty of the decision-making situation and the duration of the time interval spent on the synthesis of the solution is analyzed. The information–time uncertainty ratio is formulated and justified, reflecting the specifics of the decision-making task in conditions of uncertainty at the macroscopic level. It is shown that optimization problems of the type under consideration are affected by the Bohr complementarity principle in its macroscopic interpretation. The information-time uncertainty ratio for the quantum level of information representation and registration is derived. The conditions ensuring the possibility of registering information carried by the microstate of the system are discussed. Naturally, a constant is introduced that characterizes the relationship between the uncertainty of measuring the result of information exchange and the duration of this process at a given temperature.

Keywords: Uncertainty ratio · Decision-making · Decision quality · Semantic information · Information entropy · Thermodynamic entropy · Macroinformation · Microinformation

1   Introduction

The choice of a solution based on a set of possible alternatives is an optimization problem in which the decision-maker (DM) is looking for the most preferable compromise between the degree of adequacy of the solution and the duration of the allowable time allotted for its adoption. As shown in [1], during the synthesis of the solution, the DM evaluates at the macroscopic level the automatically arising information-time uncertainty ratio.

Note that in such tasks, as a rule, there is a specific goal, the probability of achieving which determines the value of the information [2, 3] integrated by the chosen solution. The target characteristic for the decision-maker (DM) is the quality of the decision, i.e., the degree of its compliance with the real conditions of the decision-making situation, and the role of restrictions is played by the a priori degree of uncertainty about these conditions, as well as the limit of time and resources allocated for decision-making.

These limitations are closely related, since the objectively existing time limit limits the possibilities of the DM in terms of directly generating complete and accurate information about the real situation of decision-making. On the other hand, the limit of resources, including the time resource, limits the ability of the DM to search for and/or acquire such information.

In general, the initial data for making (choosing) a decision are reliable information about the state of the system that is the control object (CO) and the state of the situation (external environment), as well as forecast data regarding possible changes in these states within an acceptable period of time. Obviously, in order to obtain such information, appropriate measurements of the essential characteristics of the control object (CO) and the situation must be carried out.

At the same time, the task arises of correctly measuring both the quality of information objects (information taken to synthesize an adequate model of the situation and choose the optimal solution) and the degree of achieving the target effect at specific costs of resources and time.

The increasing complexity of the situation model synthesized by the DM is associated with an increase in the degree of uncertainty of the situation and the risk generated by it. The set of alternatives accepted for consideration can be replenished during the selection process. In addition, the principle of choosing the optimal alternative, as a rule, also remains not formalized, which forces the DM to resort to the use of heuristic procedures.

2   Materials and Methods

In practice, the decision maker, as noted above, seeks a compromise between the degree of adequacy of the synthesized model of the real situation (completeness and accuracy of the model) and the time allotted for its generation, the synthesis of the decision and bringing to the CO the appropriate control action (decision implementation). It is obvious, that model accuracy and, therefore, the degree of solution adequacy in this case can be improved by increasing the duration of procedures for model synthesis and its transformation into a solution within an admissible time interval. Here and hereinafter, by a solution we will understand an unambiguous choice of a control action, which is to be brought to the CO.

Thus, the problem of the optimal compromise between the residual entropy (degree of uncertainty) of the adopted (chosen) decision ΔHr and the time ΔT, spent by the DM for its adoption within the permissible value ΔT_l formally can be represented as follows [1]:

$$\Delta H_r \, \Delta T \approx const; \ \Delta T \leq \Delta T_l \tag{1}$$

The residual entropy is defined as the difference between the a priori entropy H_0 and the entropy removed in the decision-making process, i.e., the information obtained ΔI:

$$\Delta H_r = H_0 - \Delta I \tag{2}$$

For further reasoning we note an extremely important circumstance. It is known that in statistical (syntactic) theory of information the meaning of a message is not considered in principle.

As for decisions causally connected with a concrete situation state, they should reflect with necessity a quite concrete sense, i.e., contain exclusively *semantic information*. Indeed, according to L. Floridi [4, 5] "the amount of semantic information in a message is determined by the degree of correspondence of the message to the real situation". Let's accept the following general definition: semantic information is a kind of syntactic information, which a physical system possesses about the environment, and which is causally necessary for the system to maintain its own existence in a state with low entropy. Hence it follows that the amount of available semantic information is a criterion of autonomy of a physical system. Let us note that syntactic information is understood here as information stated in accordance with a set of rules of ordering of words and word combinations of the used language (meaningful text).

Semantic information (information), necessary for the construction of an adequate model and the subsequent choice of the optimal decision of the DM, in principle, can be obtained in three main ways: generate yourself, find in the information environment or purchase on the market conditions. In any case, it is necessary to have sufficient resources, including time.

Resource-time in the theory of efficiency is accepted to allocate in an independent category as time and other resources, including, financial, information, computing, intellectual (knowledge), are connected heterogeneously. Since to reduce the period of time spent on achieving the goal, DM with the need to use more of these or other resources.

As a result, it is the quality of information used by DM for decision-making (primarily, their reliability, relevance, completeness and accuracy) determines the value of the decision in terms of the prospects of its successful use on purpose, i.e., for the timely implementation of the control action to transfer the control object to the next required state. As a measure of achieving the target state is efficiency, which integrates its private indicators, such as performance, efficiency and resource intensity.

So, DM is often also forced to find a compromise between ΔT time period and various resource consumption ΔR if the residual entropy ΔH_r should not exceed a certain admissible value ΔH_l:

$$\Delta R \Delta T \approx const; \quad \Delta H_d \leq \Delta H_l \tag{3}$$

Obviously, in the case when the time interval ΔT_l is long enough, resource constraints play a major role:

$$\Delta H_d \cdot \Delta R \approx const; \quad R \leq \Delta R_l \tag{4}$$

Thus, the considered optimization problems are covered by Bohr's additionality principle [6], according to which two mutually exclusive classical notions provide a complete description of the phenomenon on a quantum level.

However, it should be specified that N. Bohr, proceeding from the uncertainty principle of quantum mechanics [6, 7], has extended the additionality principle to social systems, in particular, by introducing such a pair of mutually complementary notions as reflection (analog of position) and action (analog of momentum). This pair of characteristics of conscious purposeful behavior is directly related to the decision-making tasks under consideration. Indeed, the synthesis of semantic information, reflecting the sense of a decision, is based on the use of the DM's thought procedures, aimed at refining the model of the situation and subjective evaluation (measurement) of the degree of risk, characterizing this situation. Therefore, the problem of choice is often exacerbated by doubts about the readiness to act - to bring the selected control action to the CO.

In the general case, according to the principle of additionality, the more one aspect of reality is clarified, the more uncertain becomes the associated second aspect of reality.

It is accepted that finding a verbal equivalent of this or that thought is similar to the effect of measurement on a quantum object.

As it is known, within a classical approach it is possible to adhere to the principle of determinism, which reflects action of laws of nature observed at the macroscopic level, in the assumption that residual uncertainty is caused by incompleteness and inaccuracy of knowledge, which DM (cognizing subject) has at the moment of decision making.

However, it is precisely this kind of uncorrected uncertainty, as a rule, that is characteristic of real life, when the lack of information is combined with the shortage of time allotted to make a decision and bring it to the control object.

Therefore, the macroscopic situation of decision making in conditions of incomplete certainty within the framework of the principle of additionality, can be associated directly with the uncertainty principle of quantum mechanics [7].

The meaning of the compromise relation (1) consists in the fact that the less time the DM spends on constructing a model of the situation, the greater will be its residual uncertainty. The analogy with one of Heisenberg's basic uncertainty relations (energy - time) seems quite appropriate. Indeed, let us write down this relation:

$$\Delta E * \Delta t \geq \hbar \tag{5}$$

Where ΔE is the energy measurement uncertainty, Δt is the measurement time interval, $\hbar$ is Planck's constant.

The peculiarity of relation (5) in comparison with other relations reflecting the uncertainty principle of quantum physics is that it does not use operator description of physical quantities due to principal absence of time operator.

Therefore, it seems justified to use it at macroscopic level in situations, when uncertainty is caused not by interaction of a measuring instrument with a quantum object, but by uncertainty of the DM in that semantics of a preferred solution is adequate to a real situation.

Indeed, at the experimental level of quantum physics a compromise between uncertainty (imprecision) of energy measurement and duration of its measurement process is sought. Similarly, in the framework of a decision-making problem a compromise between uncertainty (imprecision) of measurement of decision value and duration of its synthesis process is sought.

Then it is reasonable to interpret relation (1) as the information-time uncertainty relation: the less time the system spends on information exchange with the environment and/or information generation, the higher the residual uncertainty of the obtained information regarding the situation state and the decision goal, and in the general case:

$$\Delta H_I * \Delta t \geq K \tag{6}$$

Where ΔH_I is the residual information entropy reflecting the degree of semantic uncertainty of a decision, K is a constant.

Let us pay attention to an important circumstance: in the framework of the decision-making problem, we operate with macroscopic information recorded (fixed) on a material carrier. Therefore, it seems reasonable to adopt in our consideration the definition of information proposed by D.S. Chernavskiy [8], according to which information is the memorized (fixed) result of choosing one alternative from several possible and a priori equal alternatives.

Concerning the calculation of the amount of semantic information we will accept the following statements. If there are ready variants of the decision one bit of information received by DM at a macrolevel, corresponds to the result of choice between two equally probable possibilities, and further one bit is added each time, when the number of variants doubles. The amount of semantic information in the message, reflecting the result of DM thinking and stating the meaning of the decision, a priori it is reasonable to calculate by analogy with the calculation of syntactic information in the meaningful text. The fact is that the degree of correspondence of the decision to the real situation is possible to measure only a posteriori - by the result of evaluation of the degree of achievement of the target effect. A priori, however, the DM is limited to the possibility of heuristic evaluation of the probability of achieving the goal.

Therefore, ΔH_I in formula (6) will be considered as the residual information entropy, in this case the difference between the maximum a priori entropy and the entropy eliminated during the time interval Δt, i.e., generated or received from the outside information.

The ΔH_I value defines the maximum accuracy (adequacy) of semantic information reflecting the state of a decision situation, achievable in the process of its measurement of Δt duration. Accordingly, to measure (generate and transmit/receive) a definite amount of semantic information with error (uncertainty) ΔH_I, the time should not be less than $\Delta t \approx K/\Delta H_I$.

Obviously, the dimension of the input parameter is bit*s.

3 Results

To determine the value of the introduced parameter K, it is necessary to consider information processes at an elementary (microscopic) level. The exchange of information in the amount of 1 bit occurs in the process of removing the uncertainty about the outcome of the implementation of one of two discrete equiprobable events (for example, during the exchange of photons or the transition of an electron from the ground state to an excited state and vice versa) and is determined by the Hartley measure $H = \log_2 N$ (where N is the number of outcome options experience).

On the other hand, in terms of thermodynamics and statistical physics, any complex system with given macroscopic characteristics can be in one of many possible microstates and is characterized by physical (thermodynamic) entropy H_T:

$$H_T = k * ln(G) \tag{7}$$

Where k is Boltzmann's constant, G is the statistical weight of the macroscopic state, characterizing the uncertainty of the current microstate of the system, with the probability of identifying a particular microstate being inversely proportional to the statistical weight.

The exchange of energy and entropy of the system with the environment leads to a change in the statistical weight, and hence to a change in the degree of uncertainty in the microstate as well. Assuming that such a change eventually may be associated with obtaining some portion of information for a more accurate setting of a microstate or vice versa with the loss of information leading to the increase of information entropy, the following relation can be obtained from (7) according to [8]:

$$\Delta H_I \approx \Delta H_T = k * \Delta ln(G) = -k \frac{\Delta I}{log_2 e} \tag{8}$$

Where ΔH_T and ΔI are the change in the amount of thermodynamic entropy and statistical information, respectively, and the use of the logarithm of the number e is related to the possibility of transition to the natural unit of information (nat).

Here and further, it is taken into account that the acquisition of information by the system (reduction of information entropy) is accompanied by an increase of thermodynamic entropy in the environment [9].

Let us emphasize an extremely important circumstance: with the transition to the microlevel we begin to operate with microinformation, which reflects the result of the removal of uncertainty regarding a random microstate of a system in a particular fixed macrostate. It is clear that it is inappropriate to identify this information with macroinformation, in particular, due to the fact that microinformation cannot be registered, i.e., does not have the property of remembering. ndeed, reception (fixation) of information consists in translation of the system into a certain steady state, while the frequency of thermal fluctuations, which erases microinformation, is of the order of 10^{-13} s^{-1}.

However, the task of determining the value and dimensionality of the value K automatically transfers the analysis into the field of information registration at the microlevel. Therefore, we take advantage of the fact that the energy of thermal fluctuations decreases significantly when the temperature decreases, and under certain conditions they may not affect the microstates of the system. Then there appears a possibility to fix information at the microlevel. As the temperature of the system approaches absolute zero, we can talk about the transition of microinformation into an indelible state.

Now we can operate with information entropy in relation to memorized microinformation, because the registration process concerns transition of a system between stable discrete states. The emergence of a limited set of discrete states makes it possible to provide recording and storage of information at the microlevel of the system. At the same time, in contrast to microinformation, which is associated with a huge number of possible microstates, the number of stable states is incomparably small.

Therefore, we can consider that at the microlevel in a certain temperature interval (values of mean energy of particles) microstates acquire the property of registerability, which allows, in principle, to store and process meaningful information. Since quantitative relation between the thermodynamic entropy change and generation of new valuable information (macroinformation) is not seen in classical definitions [8], further we can limit ourselves to taking into account only informational entropy.

Let's compare (8) with Heisenberg uncertainty relation (5) for energy-time components. If we assume that only the information entropy of the system (and, accordingly, the information) changes as a result of the exchange with the external environment, then the relation for the change of its energy is valid:

$$\Delta E = T \Delta H_I \qquad (9)$$

Where T is the absolute temperature of the system.
Then inequality (5) will take a form:

$$T \Delta H_I * \Delta t \geq \hbar \qquad (10)$$

Therefore,

$$\Delta H_I * \Delta t \geq K = \frac{\hbar}{T} \qquad (11)$$

Substituting the value of ΔH_I from (8) into (11), we obtain:

$$T|\Delta H_I| * \Delta t = kT \frac{|\Delta I|}{log_2 e} * \Delta t \geq \hbar \qquad (12)$$

Note that in (12), in contrast to (9) and (10), the symbol |Δ...| does not mean the interval of values of quantities, but the uncertainty in their assignment, so the module of relation (8) is used. Also note that the energy of the thermal quantum kT can be interpreted as the minimum portion of energy that must be expended to obtain one nat of information. Accordingly, when receiving one bit of information, the minimum energy of the thermal quantum is kT/1.44.

As it is known [9, 10], the corresponding local decrease of the information entropy is inevitably accompanied by an increase in the thermodynamic entropy of the system as a whole, i.e., $|\Delta H_T| > \approx |\Delta H_I|$, which gives the basis for the correct use of expression (8).

Conclusively, to estimate the relation between inaccuracy in setting (obtaining) information and inaccuracy in measuring temporal characteristics, the relation is valid:

$$kT \frac{|\Delta I|}{log_2 e} * \Delta t \geq \hbar \qquad (13)$$

And further

$$|\Delta I| * \Delta t \geq log_2 e * \frac{\hbar}{k} * \frac{1}{T} \qquad (14)$$

The appearing constant κ with the dimension of bits*s*K0, has the value:

$$\kappa = log_2 e * \frac{\hbar}{k} \simeq 1.14 * 10^{-11} \qquad (15)$$

As it is easy to see, this constant is the ratio of two fundamental constants, namely Planck's constant and Boltzmann's constant. Its dimension, unlike the constant parameter K introduced at the macroscopic level of consideration, includes the absolute temperature parameter, reflecting the change of energy of particles.

4 Discussion

The obtained constant reflects the interrelation on quantum level of uncertainty (inaccuracy) of measurement of the result of exchange of fixed information and duration of this process at a given temperature.

In particular, at normal temperature ($300 \ K^0$):

$$\Delta t \geq 1.14 * 10^{-11} * \frac{1}{300 * 1} \approx 38 * 10^{-15} \tag{16}$$

Consequently, with this order of the duration of the exchange process, the error in the measurement (registration) of the amount of information will be one bit. On the other hand, for the measurement error of 1 bit of information amount at normal temperature (300 K0), the estimate of accuracy of fixing the information exchange time will be about 40 femtoseconds, which corresponds to the time of thermal fluctuations in normal conditions.

From the formal analysis (14) and (16) it follows that as the system temperature decreases, the time interval Δt required to measure the amount of information with an accuracy of 1 bit grows, i.e., the uncertainty (errors) in information transmission and recording grows.

However, in the case of temperature values close to absolute zero, the number of possible microstates in any system is significantly reduced, and they acquire the properties characteristic of macrostates in terms of information fixation. Then the temperature of the system reflects the energy of the particles participating in the information exchange.

These are the conditions necessary to ensure the possibility of recording the information. In this case the error in obtaining and fixing the information will be determined by the inaccuracy in the measurement of the number of particles at a given inaccuracy in the measurement of energy of the system ΔE.

When considering the obtained results on the phenomenological level the picture can be presented as follows.

For a very small number of equal microstates (about 10) and a small number of particles realizing them, the notion of temperature as a thermodynamic characteristic of a macroscopic system loses its meaning. Instead of system temperature it is reasonable to operate with energy of particles. The process of fixing information at the quantum level can then be described using a binary Hartley measure, considering as an elementary system, for example, the ground and excited states of an atom, provided that there is no degeneracy.

As the temperature approaches absolute zero, the number of microstates decreases and the thermodynamic entropy tends to zero. As for microinformation fixation, the unrecoverable uncertainty of energy of information exchange will be related to a number of quantum states of the system falling into an appropriate interval ΔE. Uncertainty of information then will be the number of quantum states in this energy interval (1 quantum -1 bit).

5 Conclusion

The use of the principle of additionality, which allows the extension to macroscopic processes of obtaining semantic information for managerial decision-making, allows us to connect them naturally with the information processes occurring at the micro level. Then there is an opportunity to determine the value and constants of the discussed information-time uncertainty relationship, reflecting the specifics of the optimization decision-making problem, by moving to the microlevel.

The fact that this constant is a ratio of two fundamental constants: Planck's constant and Boltzmann's constant, may indicate that the approach we develop, at least in a number of cases, does not depend on the nature of uncertainty, and it may be of interest from the point of view of further development of the theory and methodology of decision-making under conditions of risk and uncertainty.

Results of this work are also useful for estimation of ultimate capabilities of quantum computers and are also of interest for discussion of a possibility to formulate the information-time uncertainty principle, reflecting objective property of interaction of consciousness and information in the framework of universal manifestation of Bohr's additionality principle.

References

1. Klimenko, I.S., Sharapova, L.I.: Optimization of managerial decisions and information-time uncertainty ratio. Bull. Russian. New. Univ. Ser. Complex. Syst. Mod. Anal. Manage. **1**, 37–43 (2021)
2. Kharkevich, A.A.: On the value of information. Probl. Cybern. **4**, 54–60 (1960)
3. Bongard, M.M.: The Problem of Recognition. Fizmatgiz, Moscow, 320p. (1967)
4. Floridi, L.: The philosophy of information. Metaphilosophy **41**(3), 420–442 (2010)
5. Floridi, L.: Semantic Conception of Information. The Stanford Encyclopedia of Philosophy (2005)
6. Held, C.: The meaning of complementarity. Stud. Hist. Philos. Sci. **25**, 871–893 (1994)
7. Heisenberg, W.: Uber den anschaulichen inhalt der quantentheoretischen Kinematik und Mechanik. Z. Angew. Phys. **43**, 172–198 (1927)
8. Chernavsky, D.S.: Synergetics and Information. Moscow: Nauka, 105 (2001)
9. Brillouin, L.: Science and information theory. Academic Press. New York, 351p. (1962)
10. Bennett, C.H.: The thermodynamics of computation. Int. J. Theor. Phys. **21**(12), 906–940 (1982)

The Impact of Digitalization in the Economy on the Infrastructure of North and Central Asia

A. S. Nechaev[1]([✉]), O. A. Morozevich[2], O. N. Kuznetsova[3], and Bao Na[1]

[1] Irkutsk National Research Technical University, Irkutsk, Russia
nas@ex.istu.edu
[2] Belarusian State Economic University, Minsk, Republic of Belarus
[3] Irkutsk State Agrarian University named after A.A. Ezhevsky, Irkutsk, Russia

Abstract. The article analyzes how digital transformation is affecting the countries of North and Central Asia (Azerbaijan, Turkmenistan, Tajikistan, Georgia, the Russian Federation et al.). It is clear that digital solutions applied in various industries have a significant impact on productivity, employment opportunities and social well-being of the population. Private initiatives and public measures aimed to digitalize the economy have been already implemented in North and Central Asia. However, challenges in areas such as infrastructure development, digital literacy and technological competitiveness are holding back digital potential of North and Central Asia. A comparative analysis of key competitiveness indicators in the digital economy of Armenia, Kazakhstan, Kyrgyzstan, the Russian Federation and Tajikistan was conducted. The growth statistics for mobile internet access and subscribers was analyzed. The share of ICT products in exports of Armenia, Azerbaijan, the Russian Federation, Georgia, Kyrgyzstan and Kazakhstan was determined. The share of enterprises using cloud technologies was determined for the Russian Federation.

Keywords: Digitalization · Digital economy · Economy · Technological advances · Innovation · Telecommunication · Management

1 Introduction

The digital economy refers to an economy that is based on digital computing technologies. It results from billions of daily online transactions between people and companies using digital devices and data. Due to the technology-related spillover effect, its dividends are dependent on the scale of international economic and technological cooperation. The digital economy is creating a new platform for cooperation between the Silk Road Economic Belt (SREB) and the Eurasian Economic Union (EAEU) [2].

With the development of technological innovations, the digital economy has become one of the most important sectors of the national economy. In the era of globalization, digitalization of production, finance and education has become a prerequisite for the involvement of countries in more efficient global supply chains and benefits from global connectivity. World experience shows that digitalization can accelerate regional integration processes [3].

A. Gibadullin (Ed.): DITEM 2021, LNNS 432, pp. 190–201, 2022.
https://doi.org/10.1007/978-3-030-97730-6_17

Technological advances, new materials and biotechnologies provide new opportunities for social and economic growth and industrial development. Researchers and practitioners give different names to these new technologies: Industrialization 4.0, The Next Production Revolution, the Third Wave or Smart Manufacturing. Combined with other global trends such as depletion of natural resources and global warming, technological advances can change the nature of industrial production, improve productivity, provide employment opportunities and improve social well-being. The next production revolution can provide various opportunities for economic development through the optimization of production and resource management, mass individualization of products and services, automation and human-machine interaction.

The North and Central Asian countries (Azerbaijan, Kazakhstan, Turkmenistan, Tajikistan, Kyrgyzstan, Armenia, Georgia, and Uzbekistan) face challenges in implementing digital technologies for economic growth and social security. They have no laws to support the digital economy. Technology management and readiness for the latest digital advances are still under development [1].

Cheap Internet access is crucial in the development of the digital economy. In North and Central Asia, mobile Internet access is expanding as the number of fixed Internet connections is decreasing [20].

Digital transformation is a priority of the National Strategy for the Development of Information Society in the Republic of Azerbaijan for 2014–2020. Digital technologies are considered to be an important strategic vector in national policies aimed to diversify the economy. This strategy involves measures aimed to develop scientific and technological potential of Azerbaijan, to improve digital capacity and ensure cybersecurity. In 2016, Azerbaijan adopted the Strategic Roadmap for the Development of Telecommunications and Information Technologies to support the ICT sector, to improve productivity and efficiency of enterprises through the implementation of advanced digital tools and to improve the quality of public services and policy development based on actual data. The program is aimed to create an independent agency responsible for developing laws on the digital economy [10].

To enforce the Decree of the President of the Russian Federation of May 7, 2018, in 2018, the National Program "Digital Economy of the Russian Federation" was developed for 2018–2024. The Program is intended to create favorable conditions for the digital economy and digital innovations in all industries, to support the application of the latest digital advances by industries and to develop digital solutions based on big data, artificial intelligence, neurotechnology, blockchain, quantum computing, the internet of things, robotics, virtual and augmented reality [5].

Political priorities and digital transformation measures implemented by the government of Tajikistan have been specified in the National Development Strategy for the period up to 2030. According to the Program, innovative sustainable development includes three stages:

– Implementation of a new economic growth model;
– Accelerated growth by attracting investment;
– Accelerated industrialization.

Under this program, Tajikistan is seeking to improve the quality of the legal and institutional systems, develop technology industries and create an innovative economy. The transition stage should be based on more effective institutional development. It is aimed at strengthening the network of national technology parks and diversifying the economy through the development of knowledge-intensive industrial sectors [12].

In 2018, Turkmenistan adopted the Concept of Digital Economy Development for 2019–2025 to support digital transformation in all industries and public agencies. The concept is complemented by a roadmap that is aimed at creating a knowledge economy and improve economic efficiency. Turkmenistan is developing an inter-agency digital exchange system to ensure better communication across the government agencies. Turkmenistan has created the public services portal e.gov.tm to improve the quality of public services and to meet the needs of rural communities [14].

The Digital Uzbekistan 2030 program sets priorities and specifies digital transformation measures. It is aimed at creating favorable conditions for attracting foreign investment, improving the national ICT infrastructure, implementing the concept of a "smart" city, and improving digital skills. The process of creating favorable conditions for the development of digital innovation and entrepreneurship is regulated by the Strategy of Innovative Development for 2019–2021. Uzbekistan is developing a Strategy for the Development of Artificial Intelligence in 2021–2022, aimed at accelerating the pace of digital transformation and creating foundations for the next industrial revolution [6].

2   Materials and Methods

It is evident that the digital economy can organically develop in some North and Central Asian countries; however, the role of governments is vital for achieving maximum efficiency and compliance with the legal acts. In addition, the government can become a key player in the process of digital transformation and structural technological changes. Comprehensive measures can contribute to the development of the digital economy and eliminate distrust in new technologies. Another problem that is common in the region is the low level of participation in lifelong learning and low-skilled employees working with digital technologies. The governments should develop training programs aimed at improving skills of employees so that they can meet demands of the digital economy. Continuing education should be encouraged [17].

According to the Digital Implementation Index presented in the table, the public sector plays a key role in the digital transformation in Kazakhstan and the Russian Federation (0.82 and 0.84, respectively) (Table 1) [16].

Table 1 shows that for the wider implementation of digital technologies, the governments should create favorable conditions. Although the governments have set the task to expand the basic ICT infrastructure, there are a number of other initiatives to encourage the digital economy: developing and making amendments to the laws on business digitalization and implementation of new technologies [18].

In terms of digitalization of public administration in North and Central Asia, the e-government development index is higher in Kazakhstan and the Russian Federation. In Kazakhstan, e-government was created in 2006 as a platform that provides services

Table 1. Key indicators of competitiveness in digitalization in the North and Central Asian countries, 2019.

Indicator	Armenia	Kazakhstan	Kyrgyzstan	Russia	Tajikistan
Electronic participation index* (0–1)	0.57	0.84	0.69	0.92	0.39
Future-oriented government (1–7)**	3.84	4.13	3.16	3.87	4.46
Adaptability of the legal framework to digital business models*** (1–7)	4.01	4.03	3.03	3.89	3.63
Subscribers to mobile communications (per 100 inhabitants)	119.04	145.42	121.92	157.89	107.61
Subscribers to mobile broadband (per 100 inhabitants)	75.87	77.57	94.03	87.28	22.83
Subscribers to fixed broadband Internet access (per 100 inhabitants)	10.76	14.14	4.27	21.44	0.07
Subscribers to Internet access via FTTH/FTTB technology (per 100 inhabitants)	5.46	7.54	2.11	15.8	-
Internet users (% of population)	64.74	78.90	38.00	80.86	21.96
Knowledge of digital technology and computer literacy among the able-bodied population (1–7)	4.54	4.69	3.85	4.95	4.44
The growth rate of innovative enterprises (1–7)	4.25	3.58	3.27	3.74	4.16

*Notes: *The e-participation index (0 to 1) assesses the use of online services to facilitate government provision of information to citizens (electronic information exchange), stakeholder engagement (e-consulting) and participation in decision-making processes (e-decision making). **Values are provided for 2019.*

to citizens. Since then, more than six million people have used the services of the e-government portal and gained access to more than 760 electronic services. All the governments of North and Central Asian countries are making efforts to implement digital solutions [8].

3   Results

The share of mobile Internet subscribers has increased since 2000 (Fig. 1). In 2017, in Azerbaijan, Armenia and the Russian Federation, the indexes approached the average of the OECD countries, while in Turkmenistan and Georgia the growth rate of mobile Internet access was insignificant [4].

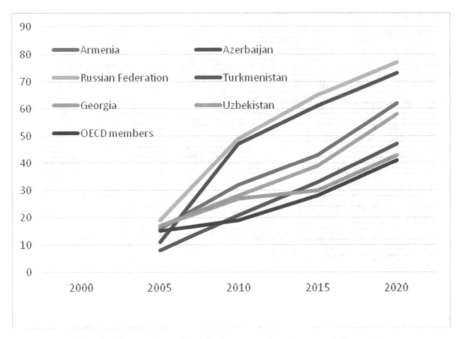

Fig. 1. The number of mobile internet subscribers per 100 people.

Limited access to the mobile Internet is a major obstacle to the development of the digital economy and e-commerce. In the North and Central Asian countries, the ICT infrastructure is based on the digital subscriber lines (DSL). Kazakhstan and the Russian Federation have begun to implement fiber-optic networks capable of providing faster Internet access. In other countries, there are no fiber-optic networks [11].

Over the last decade, most North and Central Asian countries have seen a steady increase in the number of fixed broadband subscribers (Fig. 2), including cable modem Internet connection, high-speed DSL, fiber-optic communications, and similar technologies.

Over the last decade, most North and Central Asian countries have seen a steady increase in the number of fixed broadband subscribers (Fig. 2) (cable modem Internet connection, high-speed DSL, fiber-optic communications, and similar technologies).

Figure 2 shows that the number of mobile Internet subscribers exceeds the number of fixed broadband subscribers. Mobile Internet is of great importance to the population and meets citizens' needs. However, stable and reliable fixed broadband is a critical part of the ICT infrastructure and indispensable for academia, the private sector and government agencies; therefore, to expand the fixed broadband networks, targeted investment is required.

As of 2019, in North and Central Asian exports the share of ICT products was about 0.23%, which is below the OECD average (7.1%). This low share is due to the weak diversification of economic structures, which are dependent on the export of natural resources. Armenia exports more ICT products, which may indicate the importance of

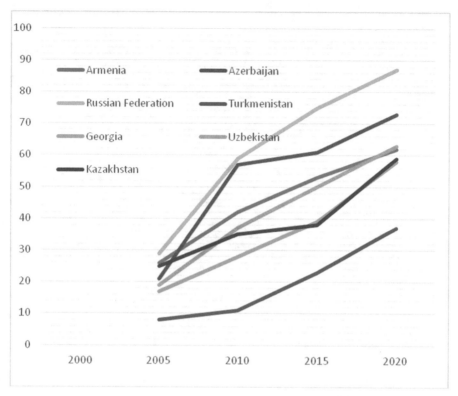

Fig. 2. The number of fixed broadband subscribers per 100 people.

the ICT sector and constraints on the growth of other industrial sectors in Armenia (Fig. 3) [9].

Data on the use of enterprise resource planning (ERP) and customer relationship management (CRM) systems can be a valuable indicator of the level of digitalization of the economy. Due to the novelty of this index, the national statistical offices of North and Central Asia do not always have these data. In 2018, in the Russian Federation with a higher level of socio-economic development, the share of enterprises that implemented the ERP and CRM systems was 21.6% and 17.6%, respectively. Presumably, in other countries, the values of this index were lower. The ERP and CRM systems are applied by North and Central Asian countries to a less extent than by the European ones. For example, in 2018, the share of Estonian companies using the ERP and CRM systems was 28% and 24%; in Finland, the share was 39%.

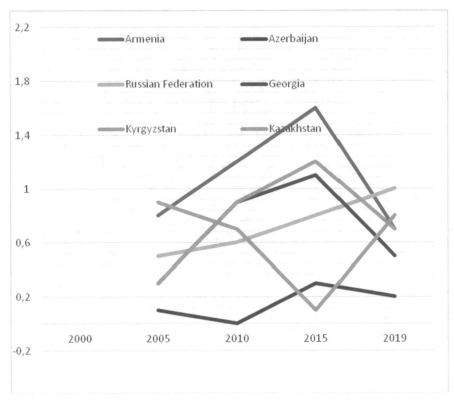

Fig. 3. The share of ICT products in exports. *Note: Data for Turkmenistan, Tajikistan and Uzbekistan are not available.*

Another index that describes the level of digitization is cloud computing. Cloud computing refers to on-demand network access to a common pool of configurable computing resources (i.e., networks, servers, repositories, applications, and services) that can be quickly provided and released with minimal management or interaction with service providers. In 2019, the share of Russian enterprises using cloud technologies was 27% (Fig. 4).

Figure 4 shows differences in the development of digital technologies in the North and Central Asian countries and the need for investment in the digital infrastructure. The developed infrastructure can be a solid foundation for economic growth and development of key economic sectors in the region. There are opportunities for regional interactions aimed to develop the regional digital infrastructure and facilitate knowledge exchange.

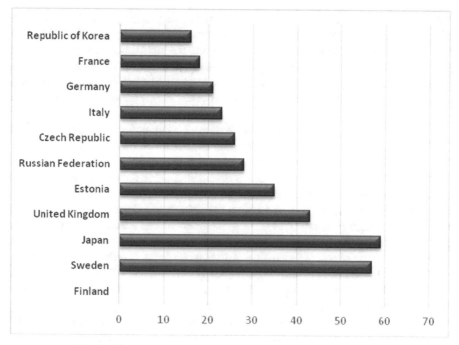

Fig. 4. The share of enterprises using cloud technologies in 2019.

4 Discussion

It is evident that the digital economy can organically develop in some North and Central Asian countries; however, the role of governments is vital for achieving maximum efficiency and compliance with the legislation. In addition, the governments can become key players in the process of digital transformation and structural technological changes. Comprehensive measures can contribute to the development of the digital economy and eliminate distrust in new technologies. Another problem that is common in the region is the low level of participation in lifelong learning and the low-skilled employees working with digital technologies. The governments should develop training programs aimed at improving skills of employees so that they can meet demands of the digital economy. Continuing education should be encouraged [13].

National statistics is indispensable for monitoring and analyzing the digital economic development. Therefore, systems and methods need to be improved and brought up to the international standards. The economic value of national statistics and accounting principles were conceptualized in the 1970s. Little has changed since then. International organizations such as the United Nations, the OECD and the World Bank have initiated discussions on how to improve the measurement of gross domestic product (GDP) in the digital era. These discussions are in their infancy and have not yet led to major changes in the international standards for statistical data collection and analysis. For example, the international community lacks reliable mechanisms for determining the value of data on economic activities and social well-being. The ESCAP Statistics Committee has developed measures aimed to improve statistical practices throughout the Asia-Pacific region. The document "Improving official statistics for implementing the 2030 Agenda for Sustainable Development: concept and measures for the Asia-Pacific Statistical Community" was adopted to produce high-quality, timely, reliable and disaggregated statistics to fill data gaps and meet the development commitments of the region [15].

Thus, updating SDG monitoring practices, even when collecting digitized data, is crucial for tracking and comparing SDG progress across North and Central Asia and developing strategies to achieve the 2030 Agenda goals.

The North and Central Asian countries should strengthen their cooperation with the international community to keep pace with the latest developments in the national statistics in the digital economy. It is advisable for the North and Central Asian countries to bring up their national statistical systems to the international standards of Eurostat and the OECD. The governments should use available resources to improve the research and analytical functions of national statistical offices, while increasing their autonomy vis-à-vis other public agencies. The growing complexity of the digital economy and the exponential growth of data force the governments to develop digital skills and competencies to apply the latest digital advances (e.g., big data, artificial intelligence).

National digital transformation strategies and programs are often based on the sectoral approach. The main areas of development of the digital economy are e-commerce, financing, agriculture and logistics. The ICT infrastructure, Internet capacity and speed, ICT infrastructure e-sustainability, the role of ICT for social sustainability are important issues [19].

The third session of the Committee on Information and Communication Technologies, Science, Technology and Innovation identified key aspects that could contribute to the recovery of the region after the COVID-19 pandemic. E-sustainability is particularly important as it is one of the four main components of the Asia-Pacific Information Highway initiative. These priorities can give long-term benefits of digital transformation to other economic and social areas.

When analyzing digital potential of North and Central Asia, it is necessary to take into account the economic structure of the region and its dependence on the agricultural sector.

The latest advances in digital technology can contribute to the restructuring of agriculture and more efficient management of agricultural resources and human capital. The agricultural sector is characterized by low productivity in all North and Central Asian countries, except for the Russian Federation.

Digital agriculture (or e-agriculture) is based on a combination of advanced technologies that allow the efficient management of resources and increase yields through the integrated monitoring and analysis of weather and soil conditions. The digital technologies and data-based approaches applied by agricultural companies have a positive impact on productivity and employment, ensure national food security and improve the living standards in rural areas.

The Ministry of Agriculture of the Russian Federation is working on a project aimed to support digital agriculture and implement new technologies in this sector. A smart planning system will be implemented throughout the country to help grow the most profitable crops and optimize supply chains. The project is aimed to implement electronic contracts with recipients of government support which will help track all export products using the paperless system.

As part of the Digital Kazakhstan program, the government has launched public initiatives aimed to improve productivity of the agricultural sector through precision farming and e-commerce. In 2018, 14 smart farms were created in Kazakhstan. Due to the data-based approaches, these enterprises have reduced their costs and doubled their productivity compared to traditional farms. Their harvest has doubled and costs have reduced by 15–20%. In 2020, Kazakhstan implemented an e-commerce platform for agricultural products to improve food security and increase export potential.

In North and Central Asia, digital agriculture is still at an early stage of development. Most countries do not have national strategies, but they have significant potential. Given the large share of the agricultural sector in GDP and its impact on the labor marker, the North and Central Asian countries need to develop legal and institutional systems and improve their digital infrastructures. The dissemination and implementation of digital technologies in agriculture should be based on public-private partnerships and close cooperation between academia and industries.

5 Conclusion

The North and Central Asian countries are heterogeneous by the level of digitalization. With the exception of Azerbaijan, Armenia, Kazakhstan and the Russian Federation, they have undeveloped communication infrastructures. The SPECA Working Group on Innovation and Technology for Sustainable Development emphasized the key role of regional cooperation and identified areas for cooperation within the SPECA strategy for developing innovation and technology. In particular, the governments have committed themselves to promoting innovation, science and technology exchange in order to ensure sustainable development in the region. Overall, the low share of R&D expenditures in GDP, weak links between businesses, universities and research organizations, and unavailable venture capital have a negative impact on the pace of digital transformation in North and Central Asia. Despite a relatively high level of education, the North and Central Asian countries lack sufficient digital skills to benefit from digital transformation. The situation has been exacerbated since there are no opportunities for lifelong learning and there is a digital divide based on gender, age and place of residence. Although the success of digital transformation depends on the availability of ICT and data professionals, knowledge dissemination mechanisms play a crucial role in realizing digital potential for the economy and society.

The national strategies and laws have been developed in the pre-digital age and need to be updated to reflect the real situation in the digital economy.

The North and Central Asian countries have already amended their laws and regulations. Due to the growing complexity of digital technologies and social and economic instability, it is necessary to develop international cooperation to identify technological problems and find new legal solutions. Legislative reforms should be aimed at improving the living standards of the population and protecting constitutional rights of citizens.

Strict legal barriers created to support innovation and dynamic business development should be avoided.

The next industrial revolution and government digitalization can improve productivity and efficiency of the public sector. To implement this policy, the North and Central Asian countries need to improve their legal and institutional systems, invest in digital skills development and improve their ICT infrastructures. Government initiatives should be based on clear priorities and contribute to the exchange of knowledge and cooperation between business and academia.

References

1. Alpatova, E., Markaryan, J., Udalov, A., Denisenko, J.: The smart city model: the concept, technology, key tasks and the prospects for the modern urbanism development. IOP Conf. Ser. Mater. Sci. Eng. **698**(7), 077019 (2019)
2. Ambika, S., Agalya, P., Sneha, K.K., Emiliya, V.R.: A study on digital India-impacts International. J. Adv. Sci. Technol. **28**(20), 2067–2073 (2019)
3. Barykina, Y.N., Chernykh, A.G.: The leasing development tools in the construction industry of the Russian Federation. IOP Conf. Ser. Earth. Environ. Sci. **751**(1), 012133 (2021)
4. Budanov, V., Aseeva, I.: Manipulative marketing technologies in new digital reality. Econ. Ann. XXI **180**(11–12), 58–68 (2019)
5. Chen, L.-S.: Design of virtual intelligent measuring and management device for charging pile based on cloud platform. Jiliang Xuebao/Acta Metrol. Sin. **40**, 116–121 (2019)
6. David, M.: Digitalization and knowledge at university: Study of collaborative student practices Digitalization of Society and Socio-political Issues 1: Digital. Communication and Culture, 169–178 (2019)
7. Ilina, E., Tyapkina, M.: Enterprise investment attractiveness evaluation method on the base of qualimetry. J. Appl. Econ. Sci. **11**(2), 302–303 (2016)
8. Khokhlova, G., Kretova, N., Sergeev, V.: Competitiveness as a factor of the company's investment attractiveness. MATEC Web. Conf. **212**, 08019 (2018)
9. Kolmykova, T., Merzlyakova, E.: Human role in the modern robotic reproduction development. Econ. Ann. XXI **180**(11–12), 183–190 (2019)
10. Koroleva, E., Sokolov, S., Makashina, I., Filatova, E.: Information technologies as a way of port activity optimization in conditions of digital economy. E3S Web. Conf. **138**, 02002 (2019)
11. Kretova, N.V., Khokhlova, G.I., Kretova, A.A., Khokhlova, A.Y.: Specificity assessment and management of financial stability of construction companies. IOP. Conf. Ser. Mater. Sci. Eng. **880**(1), 012096 (2020)
12. Nechaev, A., Antipina, O.: Tax stimulation of innovation activities enterprises. Mediterranean. J. Soc. Sci. **6**(1S2), 42–47 (2015)
13. Nechaev, A., Romanova, T., Tyapkina, M.: Author's toolkit of the state regulation of the development of leasing. MATEC Web Conf. **212**, 09010 (2018)
14. Nechaev, A.S., Zakharov, S.V., Barykina, Y.N., Vel'm, M.V., Kuznetsova, O.N.: Forming methodologies to improving the efficiency of innovative companies based on leasing tools. Journal of Sustainable Finance and Investment (2020)
15. Nurwahidah, L.S., Julianto, C.D., Sahidin, D.: Mobile learning contribution in improving the understanding of news item text generic structure. J. Phys. Conf. Ser. **1402**(7), 077–057 (2019)
16. Polozhentseva, Y., Klevtsova, M., Leontyev, E.: Effects of the economic space digitalization in the context of modern society transformation. Econ. Ann. XXI **180**(11–12), 78–87 (2019)

17. Prokopyeva, A.V., Nechaev, A.S.: Key features of risks of company innovative activities. Middle East. J. Sci. Res. **17**(2), 233–236 (2013)
18. Vivek, V., Chandrasekar, K.: Digitalization of MSMEs in India in context to industry 4.0: challenges and opportunities. Int. J. Adv. Sci. Technol. **28**(19), 937–943 (2019)
19. Yushkova, N.G., Dontsov, D.G., Gushchina, E.G.: Optimization of planning of territorial systems in the context of strategic tasks of advanced development. IOP. Conf. Ser. Mater. Sci. Eng. **698**(3), 033008 (2019)
20. Zakharov, S., Shaukalova, A.: Methodological aspects of optimization of small enterprises in modern conditions of the Russian economy. IOP. Conf. Ser. Mater. Sci. Eng. **667**(1), 012108 (2019)

Blockchain Technology as a Factor Affecting the Digitalization of the Financial Sector

Dmitry Antipin[1]([✉]), Olga Morozevich[2], Victoria Deitch[3], and Alla Gomboeva[4]

[1] Irkutsk National Research Technical University, Irkutsk, Russian Federation
dmitrii_antipin@mail.ru
[2] Belarusian State Economic University, Minsk, Republic of Belarus
[3] Irkutsk State Agrarian University named after A.A. Ezhevsky, Irkutsk, Russian Federation
[4] Buryat State Academy of Agriculture, Ulan-Ude, Russian Federation

Abstract. The current development of the global economy indicates the inefficiency of some financial mechanisms which encourages market participants to implement new technologies intended to change the financial sector and the economy as a whole by transforming business models and attracting new partners. One of the promising technologies that can have a significant impact on the financial and economic processes is blockchain. Issues affecting blockchain applications in the financial, management, logistics and other sectors and impacts of this technology on their efficiency require further research. The article aims to analyze blockchain applications in the financial sector, the dynamics of indicators describing the level of penetration of this technology in the banking sector and tools used to develop a digital infrastructure in this area. The article describes blockchain applications in credit institutions, taking into account advantages of this technology. The model of blockchain applications in the financial sector aimed at improving the use of digital technology is also described. Possible options for further development of the financial sector using blockchain are analyzed.

Keywords: Blockchain · Digitalization · Financial sector

1 Introduction

One of the main factors in the economic development is the use of digital technologies in the financial market. Digitalization covers areas and processes, which reduce savings on labor resources and all related costs, allow banks to work with customers depending on their preferences, to make banking services available, to improve the quality of banking services and increase competition in the banking sector. Digitalization has become a strategic priority for the world banking industry. The driver of changes in the financial markets are financial technologies.

The financial capabilities of banks allow them to invest in innovations. If until recently the main task of IT was to implement business goals using digital technologies, in the era of transformation, the IT departments are searching for solutions that will help change business processes [5, 7, 14].

A. Gibadullin (Ed.): DITEM 2021, LNNS 432, pp. 202–212, 2022.
https://doi.org/10.1007/978-3-030-97730-6_18

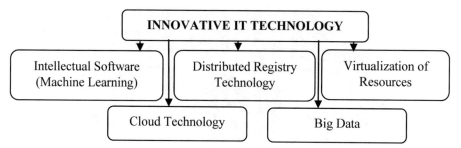

Fig. 1. The most popular innovative technologies used by banks (compiled by the authors).

The IT-architecture of banks is changing, and innovative IT-technologies are being implemented (see Fig. 1).

Innovative IT-technologies affect the ways of providing banking services, and relationships with customers.

The article deals with the most promising technology of distributed registries (blockchain).

Blockchain is a distributed and decentralized database created by participants, in which it is impossible to falsify data due to the chronological record and public confirmation by all participants in the transaction network. Its key features are algorithms for mathematical calculations and no humans in the decision-making system [19, 20]. The model of interaction of participants in the centralized system of calculations and blockchain networks is presented in Fig. 2 (see Fig. 2).

Bitcoin as the first blockchain application has contributed to the popularity of this technology. Blockchain has increased speed of exchange, reduced time costs, improved quality, reliability, availability and transparency of services and reduced operational risks.

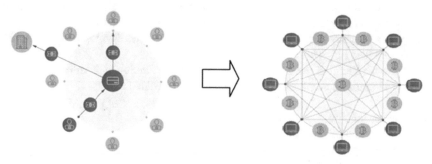

Fig. 2. The centralized payment system (left) and the blockchain network (right) [23].

Blockchain-based projects are widely implemented when providing financial, government services, real estate, notarial, transport and logistics, IoT, health, intellectual property management services as well as in the energy industry (see Fig. 3).

Fig. 3. Blockchain applications (compiled by the authors).

The main purpose of this study is to describe potential blockchain applications in the financial sector, determine the level of prevalence of blockchain-based products and develop tools aimed at improving the efficiency of blockchain applications in the financial sector.

2 Materials and Methods

New digital technologies have been discussed by governments, businesses, foreign and Russian researchers [2–22].

In the current study, the methods of analysis, comparison and classification were used.

Using statistical data, the dynamics of digital technologies, including blockchain used in the financial sector, was analyzed.

The global financial technology market is one of the fastest growing. This means that projects that seemed fantastic have been already implemented.

Figure 4 presents different types of factors affecting the development of financial technologies (see Fig. 4).

Despite the fact that the market of new financial technologies is in its infancy, it is possible to identify a number of fast-growing interconnected niches:

– Alternative lending,
– Payments and transfers (and related infrastructure),
– Asset management;
– Retail banking (Table 1).

In 2018, China (87%), India (87%) and the Russian Federation (82%) were leaders by the share of fintech users in the total number of digital technology users. In the USA, where the largest technology companies are located, the level of penetration of fintech services was only 34% (24th position in the ranking). These high positions of the emerging markets are due to the demographic factor rather than the mass use of technological finances by the population.

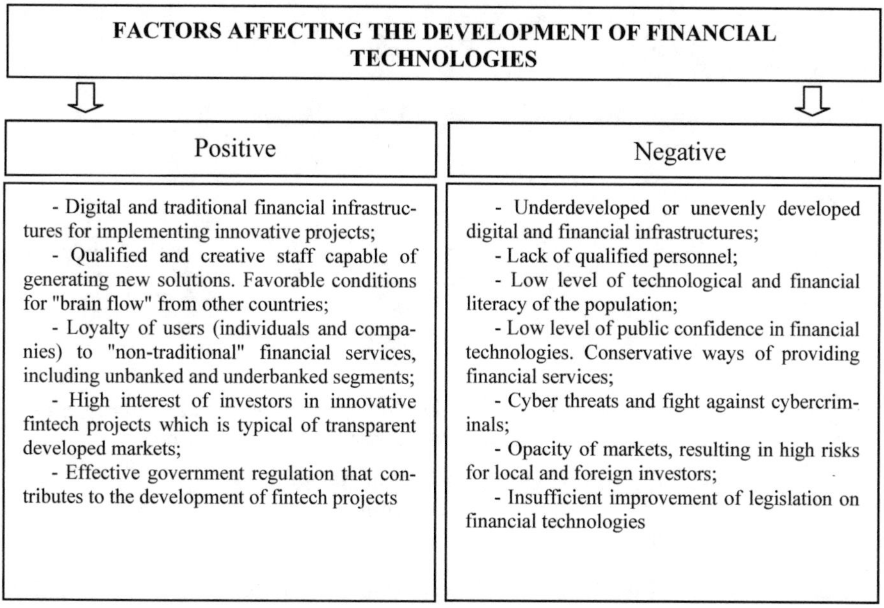

FACTORS AFFECTING THE DEVELOPMENT OF FINANCIAL TECHNOLOGIES	
Positive	**Negative**
- Digital and traditional financial infrastructures for implementing innovative projects; - Qualified and creative staff capable of generating new solutions. Favorable conditions for "brain flow" from other countries; - Loyalty of users (individuals and companies) to "non-traditional" financial services, including unbanked and underbanked segments; - High interest of investors in innovative fintech projects which is typical of transparent developed markets; - Effective government regulation that contributes to the development of fintech projects	- Underdeveloped or unevenly developed digital and financial infrastructures; - Lack of qualified personnel; - Low level of technological and financial literacy of the population; - Low level of public confidence in financial technologies. Conservative ways of providing financial services; - Cyber threats and fight against cybercriminals; - Opacity of markets, resulting in high risks for local and foreign investors; - Insufficient improvement of legislation on financial technologies

Fig. 4. Factors affecting the development of financial technologies (compiled by the authors)

Table 1. Distribution of fintech initiatives by areas (by the number of organizations) [24].

20%	Lending	7%	Crowdfunding
16%	Payments and transfers	9%	Currency transactions. Forex
12%	Asset management	5%	Safety
10%	Retail banking	14%	Other
7%	Insurance		

As for the digitalization of Russian financial companies, broadband Internet, cloud technologies, electronic data exchange technologies are gaining popularity among them (see Fig. 5).

The financial market is affected by the next generation technologies that are directly related to the Internet (mobile technologies (mobile banking), big data, artificial intelligence, virtual and augmented reality, contactless technologies, biometric technologies, distributed registry technologies (blockchain).

Thus, the use of new digital technologies in the financial sector is a relevant issue. However, there are many unresolved issues related to their use in the financial sector.

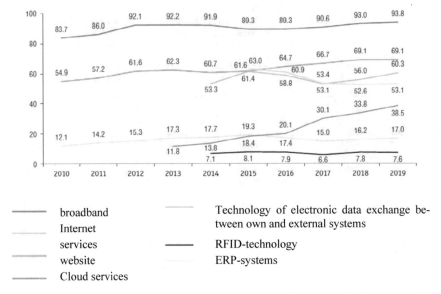

broadband
Internet
services
website
Cloud services

Technology of electronic data exchange between own and external systems

RFID-technology
ERP-systems

Fig. 5. The dynamics of digitalization of financial companies in the Russian Federation [24].

3 Results

Blockchain is one of the most controversial, but popular technologies in the financial
sector. The analysis identified a number of problems that prevent the wider use of this
technology in the financial sector (see Fig. 6).

The statistical analysis showed that in 2018, European companies spent a total of
$400 million on blockchain solutions, and almost half of the investment fell on the
financial sector. Banks, insurance, leasing and investment companies in Europe spent
more than $172 million on the development and implementation of blockchain products
(43% of the total market volume). The top three industries with the highest expenditures
on blockchain are the manufacturing and extractive industries ($80 million) and the
service sector ($44 million) (see Fig. 7).

In 2017, Morgan Stanley analysts claimed blockchain could cut bank costs by up
to 50 percent. Blockchain can optimize the infrastructure, reduce costs and increase the
RoE (return on equity) ratio.

Among its advantages are transparent payments, low transaction costs, and precon-
ditions for implementing the Industry 4.0. Its advantages outweigh the drawbacks of the
technology, which is still in its infancy and requires improvement (see Fig. 8).

The use of blockchain in the banking sector is a paradox, since it is based on decentralization, while the banking system is based on full centralization and total control. Bitcoin as a blockchain application was created as an alternative to the traditional payment
system.

The effectiveness of solving these problems depends on the implementation of a comprehensive system of blockchain applications in credit institutions, taking into account
its advantages presented in Fig. 9 (see Fig. 9).

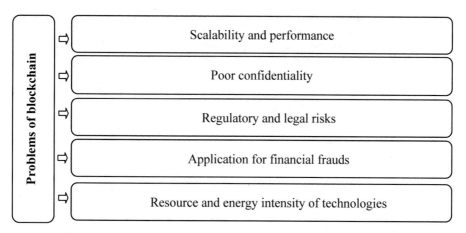

Fig. 6. Problems of blockchain technologies (compiled by the authors).

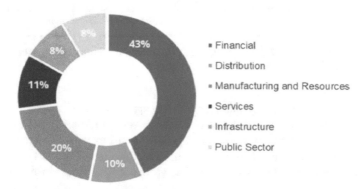

Fig. 7. European industries with the highest expenditures on blockchain [23].

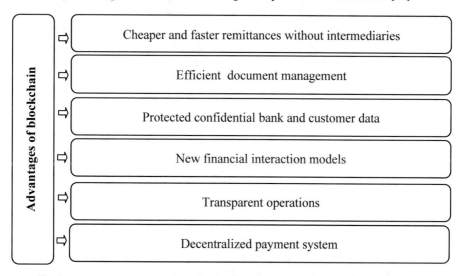

Fig. 8. Advantages of blockchain for the financial sector (compiled by the authors).

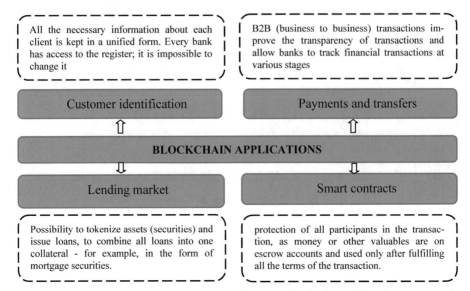

Fig. 9. The system of blockchain applications in credit institutions taking into account its advantages (compiled by the authors).

The blockchain technology can have different applications in banks.

Payments and international transfers are promising blockchain applications. The lion's share of transactions are internal and external transfers to bank customers.

The design of bank transfers is based on the addresses of at least two financial institutions with intermediaries - intermediate accounts. This slows down the financial transactions and increases costs. The commission fee can be up to 10%, and the transfer can take up to five banking days, especially if it is external. In addition, the bank deducts an additional fee when converting one currency to another.

The blockchain technology can save up to $ 12 billion a year. In particular, banks can implement the Lightning Network technology, use cryptocurrency with a low commission fee or develop a payment system with free transactions (see Table 2).

Table 2. Speed and commission fees when performing transactions with various cryptocurrencies [23].

	Bitcoin	Ripple	Ethereum	IOTA	Stellar
Speed	78 min	4 s	6 min	3 min	5 s
Commission	$0.35	$0.00003	$0.104	any	0.00001XLM

Banks can use these options or create their own ones to reduce the cost of remittances and increase the speed of transactions because they have an infrastructure, a large customer base and free financial resources to invest.

In early November 2021, Juniper Research analysts claimed that blockchain used for cross-border payments can reduce bank expenditures from $ 301 million in 2021 to $ 10 billion in 2030. This technology will make payments transparent and traceable, which is a critical advantage in the omnichannel payment market.

Client identification is one of the main functions of any financial institution. In world practice, this process is called "know your customer" (KYC). This process is often hampered by a variety of data, which are stored by other organizations. In addition, the form in which the data are provided is not ideal and cannot satisfy all stakeholders. For financial companies, the average KYC cost is $48 million, for banks - $70 million, for companies with an income of more than $10 billion – more than $ 150 million. The KYC procedure takes 26 days, and users must go through it every time they "meet" a new financial company or bank.

The blockchain technology can become the gold standard for storing users' financial and personal data, including information about the origin of money, business interests, history of financial transactions, etc. Banks and other financial institutions will be able to use this information to expedite the KYC procedure and keep the certificate, which will be valid for subsequent inspections (see Fig. 10).

One of the most popular blockchain applications in the financial sector is data security. In 2018, financial companies lost $ 3.8 billion due to personal and business data leaks (Fig. 11). The average value of each stolen or lost record that contains personal and/or confidential data was $148.

Companies providing financial services are victims of cyberattacks 300 times more often than companies in other industries, as it is easiest to get direct access to users' money (see Fig. 11).

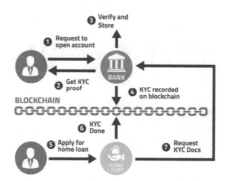

Fig. 10. KYC blockchain implementation using the certificate [23].

As a result, banks and other financial companies are forced to invest larger sums in ensuring the security and reliability of their services.

Blockchain-based security and authentication tools are more likely to detect and block malicious attempts aimed to obtain data through black holes. As a result, banks and insurance companies can reduce data leakage costs and raise the level of customer confidence.

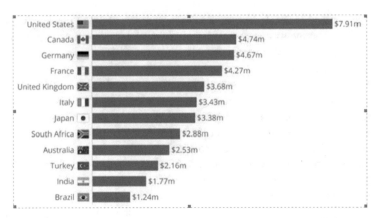

Fig. 11. Average cost of data leakages by country in 2018 [23].

There are other blockchain applications. The technology can be used in bank areas such as asset valuation and verification, counterparty risk audit, operational risk reduction, etc.

The analysis conducted made it possible to develop a model of blockchain applications in the financial sector (see Fig. 12).

The model can be used to determine options for solving the problems that arise in the implementation of the blockchain technology.

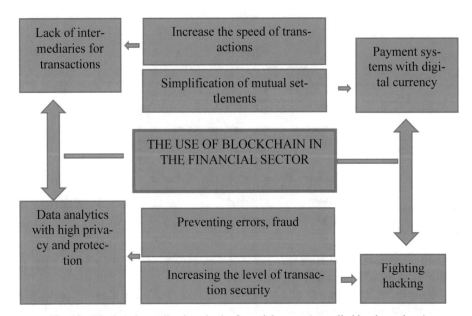

Fig. 12. Blockchain applications in the financial sector (compiled by the authors).

4   Discussion

Thus, blockchain is one of the most promising technologies in the financial market, but to implement and legalize it, risks should be analyzed and laws should be adopted. Although commercial banks have succeeded in developing digital business models, enabling millions of people to use Internet banking and becoming experts in providing data-driven services, the process of implementing blockchain in the financial sector is a challenging task [1–16].

Further research in this area may be aimed at developing recommendations for the legislative consolidation of various issues, such as the status of cryptocurrency. Secondly, it is necessary to settle transactions on cryptocurrencies. Third, it is necessary to regulate the ICO.

Addressing these issues will allow countries to create an efficient blockchain infrastructure, which will contribute to the use of blockchain technologies in various areas, accelerate the digitalization of the economy and create a digital society.

5   Conclusion

The number of areas where blockchain is applied is increasing. As compared to other countries, in Russia, this technology is less popular.

The model of blockchain applications in the financial sector and the system of blockchain applications in credit institutions are aimed at improving the efficiency of this digital technology in these areas.

Technological financial products and services are gaining popularity in various traditional segments. According to a number of studies conducted by analytical companies, in the Russian Federation the most promising areas are payments and transfers, financing, capital management and insurance. At the same time, blockchain is to become one of the key elements in the digitalization of the economy and development of a digital society. This will optimize processes in many areas and make the financial sector more transparent and efficient.

References

1. Antipina, O.V., Velm, M.V.: Characteristics of project management in the construction industry of the Russian Federation in modern economic conditions. IOP Conf. Ser. Earth Environ. Sci. **751**(1), 012072 (2021)
2. Barykina, Y.N., Chernykh, A.G.: The leasing development tools in the construction industry of the Russian Federation. IOP Conf. Ser. Earth Environ. Sci. **751**(1), 012133 (2021)
3. Barykina, Y.: Analysis of information support for innovation development. IOP Conf. Ser. Earth Environ. Sci. **667**(1), 012012 (2019)
4. Boreiko, D., Ferrarini, G., Giudici, P.: Blockchain startups and prospectus regulation. Euro. Bus. Organ. Law Rev. **20**(4), 665–694 (2019). https://doi.org/10.1007/s40804-019-00168-6
5. Cai, C.W.: Disruption of financial intermediation by FinTech: a review on crowdfunding and blockchain. Account. Finan. **58**(4), 965–992 (2018)
6. Feng, L., Zhang, H., Tsai, W.-T., Sun, S.: System architecture for high-performance permissioned blockchains. Front. Comput. Sci. **13**(6), 1151–1165 (2019)

7. Fernandez-Vazquez, S., Rosillo, R., De La Fuente, D., Priore, P.: Blockchain in FinTech: a mapping study. Sustainability **11**(22) (2019)
8. Kabakova, O., Plaksenkov, E.: Analysis of factors affecting financial inclusion: ecosystem view. J. Bus. Res. **89**, 198–205 (2018)
9. Kotishwar, A.: Impact of digitalization on select banks. Indian J. Finan. **12**(12), 32–51 (2018)
10. Kuklina M.V., Trufanov A.I.: Network platform for tourism sector: transformation and interpretation of multifaceted data. Sustainability **12**(16), 6314 (2020)
11. Kuklina M.V., et al.: Land use in remote areas: socio-economic prospects. IOP Conf. Ser. Earth Environ. Sci. **885**, 012030 (2021)
12. Kuznetsov, S.M., Gavrilova, Z.L., Teplouhov, O.J., Avetisyan, B.R., Markova, S.V.: Improving the methods of monitoring and automation and mathematical modelling of railway protection. J. Phys. Conf. Ser. **1333**(4), 042021 (2019)
13. Liu, Q., Zou, X.: Research on trust mechanism of cooperation innovation with big data processing based on blockchain. EURASIP J. Wirel. Commun. Netw. **2019**(1), 1–11 (2019). https://doi.org/10.1186/s13638-019-1340-5
14. Majuri, Y.: Overcoming economic stagnation in low-income communities with programmable money. J. Risk Finan. **20**(5), 594–610 (2019)
15. Nechaev, A.S., Antipin, D.A.: Mechanism for assessing the efficiency of financing the enterprise innovative activities. Actual Prob. Econ. **154**(4), 233–237 (2014)
16. Nechaev, A., Antipina, O.: Taxation in Russia: analysis and trends. Econ. Ann.-XXI **1–2**(1), 73–77 (2014)
17. Nechaev, A., Prokopyeva, A.: Identification and management of the enterprises innovative activity risks. Econ. Ann.-XXI **5–6**, 72–77 (2014)
18. Porsev, E.G., Gavrilova, Zh.L.: Modern approaches to water drying in the underground transport system. IOP Conf. Ser. Mater. Sci. Eng. **560**(1), 012196 (2019)
19. Shan, J.-Y., Gao, S.: Research progress on theory of blockchains. J. Cryptol. Res. **5**(5), 484–500 (2018)
20. Silva Filho, A.C., Maganini, N.D., de Almeida, E.F.: Multifractal analysis of Bitcoin market. Phys. A **512**, 954–967 (2018)
21. Zakharov, S.V., Troshina, A.O., Lobova, A.U.: The method of selection of innovative solutions based on an assessment of the efficiency reserves. In: Proceedings of the 2017 International conference "Quality Management, Transport and Information Security, Information Technologies", IT and QM and IS 2017, pp. 601–602 (2017)
22. Zakharov, S., Shaukalova, A.: Methodological aspects of optimization of small enterprises in modern conditions of the Russian economy. IOP Conf. Ser. Mater. Sci. Eng. **667**(1), 012108 (2019)
23. Merehead. Homepage. https://merehead.com/ru/blog/how-use-blockchain-banking-use-cases/. Accessed on 14 Nov 2021
24. Dcenter. https://dcenter.hse.ru. Accessed on 10 Nov 2021

New Trends and Digital Models of E-commerce in Era of COVID19: Case Study of Russian and Korean Retail Companies

Olga Shvetsova[1](✉) (iD), Anna Kuzmina[2] (iD), and Victoria Levina[2] (iD)

[1] Korea University of Technology and Education, Cheonan City 31254, South Korea
`shvetsova@koreatech.ac.kr`
[2] State Electrotechnical University LETI, Saint-Petersburg 197376, Russian Federation

Abstract. This paper presents recent survey of e-commerce trends and its digital models in Russian and Korean markets, especially in the field of customization. This article is based on 2 types of research: survey and case study. The survey includes statistical data and it has been applied to separate domains of e-commerce in Russia and South Korea at the same industry. Authors discuss retail industry for B2C market. The period of survey is 2021. There are 20 small and medium size companies in Russian market and 15 small companies in Korean market were investigated. The paper discusses trends in the development of the international e-commerce, ways of raising revenue from e-commerce and changes that took place in the period of pandemic. The development of e-commerce has the potential to make a significant contribution to economic growth and social welfare through increased consumption households, expansion of the retail sector and a decrease in operating costs. In contrast to the linear model of innovation, when the cost of innovation is gradually transformed into the results of innovation, in the digital economy, various types of innovative activities develop in parallel to each other and are characterized by continuous improvements through customer feedback.

Keywords: Digital model · E-commerce · COVID19 · Digital technology

1 Introduction

A combination of technological advances from advances in digital technology to creation of new materials and biotechnology opens up new opportunities for socio-economic growth and industrial development. Thanks to researchers and practitioners the results of the application of new technologies have received such names as Industry 4.0, the Next Manufacturing Revolution, the Third Wave, or Smart Manufacturing. In combination with other global trends such as natural resource depletion and global warming, technological advances will change the nature of industrial production, having a huge impact on productivity, employability and well-being of society. The next industrial revolution opens up a variety of opportunities for economic development through the optimization of processes and use resources, mass individualization of products and services, industrial automation production and human-machine interaction [1].

A. Gibadullin (Ed.): DITEM 2021, LNNS 432, pp. 213–223, 2022.
https://doi.org/10.1007/978-3-030-97730-6_19

Digital technologies, including artificial intelligence and big data, are general purpose technologies that have "a range of characteristics [and] especially well-suited for providing longer term enhancement productivity and growth across a wide range of industries" [1]. By compared with previous stages of industrial development current restructuring industrial production is characterized by higher instability and speed changes.

Electronic commerce is the sale, purchase, delivery of orders, or the assumption of an obligation to ordering products, goods and services over the Internet [2, 3]. To denote online transactions with intangible digital solutions regulators and researchers use the term digital commerce, while e-commerce transactions trade associated with traditional physical products [5, 6]. According to UNCTAD (2018), Global e-commerce sales (to business-to-business and business-to-consumer) reached $26 trillion in 2018. US dollars, which is 30% of world GDP [7].

The development of e-commerce has the potential to make a significant contribution to economic growth and social welfare through increased consumption households, expansion of the retail sector and a decrease in operating costs [8].

In contrast to the linear model of innovation, when the cost of innovation is gradually transformed into the results of innovation, in the digital economy, various types of innovative activities develop in parallel to each other and are characterized by continuous improvements through customer feedback. Growth in partnerships across industries and networking indicates the need to pool resources to investment in R&D and combining scarce skills and competencies in the field of working with digital technologies. The most important competitive advantage of the digital economy is the ability to capitalize on open innovation and crowdsourcing opportunities [9]. It requires changes in the organizational culture to overcome resistance using solutions developed externally.

The impact of digital technology and data-driven approaches on enterprises is highly versatile and includes shorter innovation cycles, massive personalization, production on demand and optimization of development processes, distribution and maintenance. Virtual simulations, digital twins, and 3D printing can enhance experimentation and optimize innovative processes. Digital products and services can be released in early alpha stages with subsequent updates and improvements based on customer feedback and analysis usage trends. Access to extensive and timely data enables more effectively predict equipment maintenance [10].

Advanced predictive models and machine learning technologies can predict a malfunction before it occurs, thereby contributing to significant increase efficiency and create value while preventing malfunction. The proliferation of smart and connected devices and the improvement of algorithms allow you to achieve significant shifts in the work of the enterprise. Manufacturing enterprises are increasingly engaged in the development of digital value-added services that would improve the functionality of their products and create new revenue streams [11, 12]. Blurring the boundaries between products and services, often referred to in academic literature as "digital service" allows carry out transformational changes in the way economic value is created [13].

2   Materials and Methods

2.1   Material

For the purpose of writing the article the following data was collected and analyzed:

1) Qualitative data

   The information presents characteristics that are not measured in numbers, while the observations themselves can be divided into the measured number of groups. The information stored in this type of variable is difficult to measure, and measurements can be subjective.

2) Quantitative data

   Information is recorded in the form of numbers and represents an objective measurement or count. Temperature, weight, number of transactions are examples of quantitative data. Analysts also call such data numerical.

Data of that types is collected by the authors from the below-mentioned books, articles and statistics and by the means of descriptive, exploratory, inferentiality, predictive, causal and mechanistic ways of analysis.

Trends in e-commerce have emerged as an issue for research in the last 10–15 years.

In the book «Trends in E-business, E-services, and E-commerce: Impact of Technology on Goods, Services, and Business Transactions» printed in USA, IGI Global, 2013 the issues of scaling e-commerce projects by franchising (through creating soft landings) are discussed, improvement of marketing performance by personalized recommendation, online corporate reputation and etc. [14].

Some authors concentrate on the analysis of e-commerce trends of the certain countries, like Richter, Christina in the book «E-Commerce Trends in China: Social Commerce, Live-Streaming oder New Retail» printed in Germany, «An Overview of E-commerce Trends in Germany» and «An Overview of E-commerce Trends in France», printed in Canada [15, 16].

According to trends, e-commerce market is developing rapidly, and the Chinese market shows the fastest growth (Fig. 1).

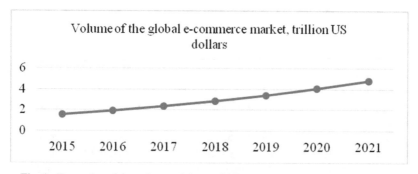

Fig. 1. Dynamics of the volume of the world E-commerce market, $ trillion [16].

During the pandemic financial center Skolkovo-RES conducted a study of more than 1,000 Russian consumers of e-commerce and, according to its results, identified 6 main trends in consumer behavior that either gained a foothold in connection with the coronavirus pandemic or appeared against the background of unexpected changes in the lives of Russians.

1. Transition from free consumption to thrift: a decrease in incomes and savings.
   The forced self-isolation regime introduced an unplanned decrease in incomes of the population. According to NAFI research, 46% of all Russians faced financial difficulties and lack of funds during the pandemic. Against the background of declining incomes, the saving behavior of Russians using the Internet has also changed. A third of those who saved money before the pandemic were unable to do so during the self-isolation period (April–June 2020). Moreover, 41% refused to buy for themselves. Some due to reduced income (57%), some due to savings (50%), and some due to the inability to get a number of goods and services online (35%) [17].
2. Promotions and bonuses continue to play an important role.
   Bonuses, points, miles and other incentive programs are additional financial opportunities. Members of loyalty programs preferred to accumulate points or save cashback in order to make significant purchases, while unplanned. Despite the fact that consumers see that not all loyalty programs are profitable and have a good conversion, they are still emotionally attached to such offers.
   Many of the consumers choose promotions and bonuses as a saving strategy and cannot imagine their usual consumption without them. According to the latest data, 49% of customers try to make purchases only if there are promotions, bonuses or discounts, and 70% are participants in various loyalty programs from banks [17].
3. Focus on Necessary and Conscious Consumption.
   The conscious consumption trend began a couple of years ago at the intersection with the trend towards sustainable consumption in order to reduce the footprint of human activity and the impact on the environment. However, the pandemic has brought a new connotation to the term. Suddenly, Russians were locked in their apartments or houses, and some of the expenses that previously seemed necessary were superfluous. 63% of Internet users noted that they began to spend money more consciously, deliberately.
   At the same time, spending on clothes and shoes (43%), hobbies and entertainment (43%) decreased. Conversely, Russians began to spend more on categories such as food and essential goods (25%) and medicines (21%) [17].
4. Living Online.
   Despite the fact that the level of Internet penetration in Russia fluctuates around 78% as of 2019, according to a study by Yandex and GfK, the share of Russians using e-commerce was 42%. According to our study, among Internet users of the Russian representative sample during the period of self-isolation, 85% made purchases online.
   By the way, those who use online commerce are mostly satisfied with the services offered - 40% of respondents say that they have not encountered problems recently. Those who experienced inconvenience when ordering online, note among the main

problems the delayed delivery time (38%) and the lack of the necessary goods (33%) [17, 18].
5. Online is cheaper than offline.
   Part of the preference to receive some services online is due to the fact that they are perceived as cheaper. 70% of all those who took part in online activities did not pay for any of them. 79% of Russians believe that online services should cost less than similar offline services. This makes online commerce a space of high expectations with low prices [18].
6. Reluctance to Change.
   The latest trend is the rejection and reluctance of change among the majority of the population. In Russia: 82% of Internet users plan to return to their usual way of life, when all restrictions imposed due to the coronavirus will be lifted. However, the longer the restrictions persist, the more Russians can get used to the new normal and move away from old habits.

2.2 Methods

This article is based on 2 types of research: survey and case study. The survey includes statistical data and it has been applied to separate domains of e-commerce in Russia and South Korea at the same industry. Authors discuss retail industry for B2C market. The period of survey is 2021. There are 20 small and medium size companies in Russian market and 15 small companies in Korean market were investigated. Description of survey participants is presented in Table 1.

Case studies have a very narrow focus which results in detailed descriptive data which is unique to the case(s) studied.

Table 1. Characteristics of research participants.

Type of company	Business characteristics	Number of participants	Number of employees	Year of establishment
Korean small-size company	Fast food delivery service, online shops	5	4–50	2010–2018
Korean medium-size company	Fast food delivery service, menu production, cooking training, online shops	10	100–150	2015–2020
Russian small-size company	Fast food delivery service, online shops	8	10–50	2015–2019
Russian medium-size company	Fast food delivery service, menu production, cooking training, online shops	12	150–200	2011–2019

3   Data Collection and Results

3.1   Case 1. E-commerce Digital Model in Russia

There are major statistical trends in e-commerce in Russian market [19]:

- Revenue in the eCommerce market is projected to reach US $28,413m at the end of 2021;
- Revenue is expected to show an annual growth rate (CAGR 2021–2025) of 5.17%, resulting in a projected market volume of US$34,766m by 2025;
- The market's largest segment is Electronics & Media with a projected market volume of US$8,009m in 2021;
- With a projected market volume of US$1,542,551m in 2021, most revenue is generated in China;
- In the eCommerce market, the number of users is expected to amount to 73.1m users by 2025;
- User penetration is about 45.3% in 2021 and is expected to hit 50.4% by 2025;
- The average revenue per user (ARPU) is expected to amount to US$430.07;
- In 2021, 94% of total eCommerce purchases in the country are domestic;
- In 2021, 52% of total eCommerce purchases are paid by Cards.

Meanwhile, there is no doubt that digitalization is already a game changer in the Russian market and that it will continue to do so, as some companies in Russia are reporting double- or triple-digit growth rates in e-commerce sales per year. For some, e-commerce has started to drive the whole business model while regular retail is growing only marginally – a transformation (or revolution) that has taken place only over the last few years.

Looking at other geographies and the pace at which they keep embracing the e-commerce ecosystem, the disruption of traditional business models seems inevitable. On a global stage, the share of e-commerce in total Russian retail is still 3–4 times smaller than in the leading digital economies, out of which China holds the top position. However, Russia keeps growing in scale and importance – especially in Moscow, which is way ahead of other regions in the country. Regardless of the challenges that digital transformation and its successful implementation entails, it seems that embracing it in a well-planned and comprehensive manner would be essential for Russia in order to be a globally-competitive market and stay so in an ever faster changing world.

Digital transformations of retail in Russia can be approached through the concept of "liquid retail," that is an open metaphor that helps to problematize the dynamics of retail, with the purpose of shedding light on the current circumstances in which retail stakeholders and consumers navigate the accelerated transformations [20]. Following this framework, in this section we pay attention to macro-level changes of market transformations, normative shifts, techno-economic (infra)structures and meso- and micro-level activities of multiple actors (new retail formations, changing consumption practices, etc.) in order to reflect upon how digitalization has changed retail and shopping in Russia (Fig. 2).

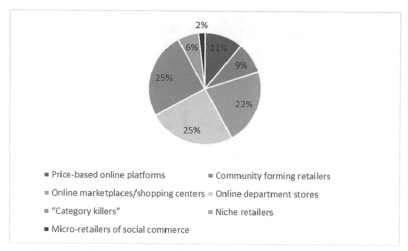

Fig. 2. Online retail platforms in Russia. Source: made by authors

3.2   Case 2. E-commerce Digital Model in South Korea

The e-commerce market in South Korea at the end of 2021 reached $71.7 billion and took a share of 24% of all retail in the country. If by volume the South Korean online retail market is considered only the fifth in the world (with quite real prospects to become the third soon), then its share in the structure of the entire retail is the highest on the planet. For the second time since 2018, China lagged behind in this indicator: the share of e-commerce in the country was 0.3% less, only 23.7% (Figs. 3 and 4).

Fig. 3. E-commerce value in South Korea [19, 20].

Fig. 4. Top online retailers in South Korea, volume of sales $ mln [19, 20].

When ordering goods on the online platform, in 99.8 cases out of 100, delivery will take no more than a day, and most likely no more than a few hours. Moreover, before bringing the parcel home, the online delivery service will first ask by phone whether it is possible to ring the doorbell or it is better to knock quietly, suddenly someone is planning to rest in the house at this time.

If the recipient is not at home, the courier will coordinate by phone with the buyer where and under what geraniums to leave his order so that he can pick up the parcel upon return. Well, if the product for some reason did not suit the consumer, then to return it is enough to leave the packaging outside the door and inform the site administration about it: no copies of the passport, statements, checks and other bureaucracy are required.

The achieved online consumer paradise has allowed leading online retail company Coupang to actively sell such seemingly inappropriate product categories as food, clothing and shoes. Yes, almost everything, the risk of being left with an unsuitable product in this country is reduced to almost zero. In this regard, Korean experts expect fashion to become the main category of online sales in the country in 2020–2021.

South Korea is one of the biggest e-commerce markets worldwide. As a consumer society with an internet usage rate of close to 92 percent among the population, it is no surprise that e-commerce holds a strong position in South Korea. Industry figures suggest that the retail e-commerce volume will continue to increase, especially due to the impact of COVID-19. Internet shopping was especially popular among people under 40, with over 94 percent of South Koreans from 20 to 39 using internet shopping [20].

As of 2020, Korea was ranked 5th in e-commerce market size, with its mobile shopping growing at the fastest pace in the world. In 2020, online shopping transactions increased even more due to the prolonged COVID-19 pandemic. In addition, the accelerating digital transformation across all economic sectors has brought about innovation in the retail industry with the application of 4th industrial revolution technologies. An evolution towards the era of Retail 4.0 is underway, spurred by the spread of the "omni-channel" (integrating online and offline channels), and the emergence of new business models based on data.

4   Discussion

As for the influence of COVID-19 pandemic to the e-commerce systems of different countries it seems obvious that the pandemic has pushed the explosive growth of e-comm segment. David Souter and Anri van der Spuy stay in their article that COVID has accelerated the upward trend [21]. It's seen economic activity as a whole turn down, domestically and internationally, but the share made up of digital transactions has increased in every region.

Usually commercial enterprise operates in conditions of increased competition and global effects of the world economic crisis. Limited financial resources, instability of the economic environment, escalating demands of consumers for the quality of services make it necessary to search for additional sales channels. With the popularization of the Internet and integrated automation of economic sectors, the role of e-commerce is becoming increasingly important. Since 2015, the B2C e-commerce turnover in Russia increased by 6.6% compared to the previous years and amounted to 21.621 million euros. Russia was ranked first in Europe in terms of the number of e-shoppers (30 million people in 2015), and this number keeps on growing steadily.

E-shopping's enabled some businesses with the right resources to maintain or expand sales, and consumers with the right resources to maintain purchases and lifestyles, during the pandemic. But the impact isn't equal everywhere. Digitalization's increased efficiency and enabled those with the right resources to build businesses or maintain their quality of life, but it's also concentrated the power of some large businesses – there've been big winners, including Amazon – and emphasized the significance of inequalities in income, connectivity, capabilities and governance [22].

The pandemic boosted sales in B2C segment and an increase in the volume of e-commerce between businesses (B2B). The increase in consumer sales is particularly noticeable in the segments of sales of medical products, household goods, basic necessities and food.

The Internet and mobile data services are experiencing a period of increased demand, which has entailed the need for their urgent technical and regulatory adaptation. However, a certain segment of online shopping experienced a sharp decline in demand: for obvious reasons, platforms for booking accommodation and air tickets recorded a record decrease in traffic.

E-commerce in goods and services has been negatively affected by the same factors that led to disruptions in supply and demand in general – logistical delays or cancellations of orders. Other problems include unreasonable price increases, product safety issues, cybersecurity and fraud issues, etc. [23].

5   Conclusion

The pandemic has highlighted the digital divide within and between countries. The need to bridge this gap in the level of digitalization of countries has become apparent. In times of crisis, access to people and businesses to quality services in the field of information and communication technologies, which includes e-commerce, telecommunications, computer and other IT services, becomes critical.

As it is mentioned in OECD research on the e-commerce perspectives during the pandemic, e-commerce allows individuals to self-isolate while retaining access to a full range of products. Whereas previously e-commerce focused on high-tech goods, toys or books, it now increasingly includes essential goods. E-commerce has enabled permanent access to the social and cultural sphere, providing the possibility of distributing tickets with time stamps of visitation to maintain distance. For many firms, e-commerce is now an important alternative or additional sales strategy to keep going despite current restrictions.

The crisis has affected the categories of demand for goods and services. Increased demand was observed for personal protective equipment, household goods, food, while goods for travel, sports, business-style clothing were not of interest. The growing demand for food has led many farmers to switch to e-commerce to ensure that products are delivered directly to the consumer. In Germany, online sales of medicines and food have grown significantly. In Korea, significant growth was observed in the field of public catering (66.3%), household goods (48%), food and beverages (46.7%), and the provision of cultural and entertainment services or the organization of travel and transportation decreased significantly. In China, food has become the single largest leader in e-commerce.

Online merchants have faced the same difficulties as traditional conventional retailers, namely reducing consumer spending on non-essential goods. The COVID-19 crisis has led to a redirection of demand from small and specialty stores to larger and more diversified ones. The complementarity of online and offline sales channels has become apparent. South Korea showed impressive cases of online-buying and increase in consumption, so the experience can be analyzed and taken into consideration by other countries.

Russian Federation made huge steps to the digitalization of different spheres, e.g. medicine, civil service. This experience can fully be used in another spheres of life – education, law, telecommunication etc.

Enterprises are now obliged to pass through digital transformation and that's the best investment to their stability.

References

1. Holmström, J., Partanen, J.: Digital manufacturing-driven transformations of service supply chains for complex products. Supp. Chain Manage.: Int. J. 9(4), 56–96 (2014)
2. Kervenoael, R., Bajde, D., Schwob, A.: Liquid retail: cultural perspectives on marketplace transformation. Consump. Mark. Cult. J. 21(5), 417–422 (2018)
3. Lu, Y.: Industry 4.0: A survey on technologies, applications and open research issues. J. Ind. Inform. Integr. 6, 1–10 (2017)
4. Global retail e-commerce sales 2014–2021. https://www.statista.com/statistics/379046/worldwide-retail-e-commerce-sales/. Accessed on 01 Nov 2021
5. e-Commerce. https://belretail.by/article/esommerce-prodaji-v-mirepo-itogam-goda-vyiroslina-do-mlrd. Accessed on 01 Nov 2021
6. Trends in E-business: E-services, and E-commerce: Impact of Technology on Goods, Services, and Business Transactions». IGI Global, USA (2013)
7. Richter, C.: E-Commerce Trends in China: Social Commerce, Live-Streaming order New Retail. Germany. Springer Fachmedien Wiesbaden (2021)
8. An Overview of E-commerce Trends in Germany. Canada, Market Access Secretariat (2017)
9. An Overview of E-commerce Trends in France. Canada, Market Access Secretariat (2017)
10. Roy, R., et al.: The servitization of manufacturing. J. Manuf. Technol. Manag. 8, 45–56 (2019)
11. Shvetsova, O.A., Lee, S.-K.: Living labs in university-industry cooperation as a part of innovation ecosystem: case study of South Korea. Sustain. J. 13, 57–93 (2021)
12. Starostin, V., Chernova, V.: E-commerce development in Russia: trends and prospects. J. Internet Bank. Commer. 21 (2016)
13. Susto, G.A., Schirru, A., Pampuri, S., McLoone, S., Beghi, A.: Machine learning for predictive maintenance: a multiple classifier approach. IEEE Trans. Indust. Inf. 11(3), 812–820 (2015)

14. Tronvoll, B., Sklyar, A., Sörhammar, D., Kowalkowski, C.: Transformational shifts through digital servitization. Ind. Mark. Manage. **2**, 12–25 (2020)
15. OECD Homepage. https://www.oecd.org/economy/surveys/. Accessed on 11 Nov 2021
16. Statista Homepage. https://www.statista.com/outlook/dmo/ecommerce/russia/. Accessed on 15 Nov 2021
17. Euromonitor International homepage. https://www.euromonitor.com/. Accessed on 12 Oct 2021
18. Global Data homepage. https://www.globaldata.com/. Accessed on 12 Oct 2021
19. Shvetsova, O.A.: New forms of effective collaboration: how to enhance big data for innovative ideas in the online environment. In: Big Data for Entrepreneurship and Sustainable Development, 1st edn. Routledge CRS press (Francis&Taylor Group), USA (2021)
20. E-commerce, trade and the COVID-19 pandemic. https://www.wto.org/english//tratop_e/covid19_e/ecommerce_report_e.pdf. Accessed on 12 Oct 2021
21. Souter, D., Spuy, A.: Inside the Digital Society: COVID-19 and e-commerce. https://www.apc.org/en/blog/inside-digital-society-covid-19-and-e-commerce. Accessed on 12 Oct 2021
22. Online, L.: How the pandemic has changed the attitude of Russians to e-commerce. Retail Loyalty **7**(94), 56–78 (2020)
23. World Trade Organization homepage. https://www.wto.org/. Accessed on 12 Oct 2021

Digitalization of Regional Economies in the Context of Innovative Development of the Country

Oksana Antipina[1](✉), Elena Kireeva[2], Natalya Ilyashevich[3], and Olga Odoeva[4]

[1] Irkutsk National Research Technical University, Irkutsk, Russian Federation
antipina_oksana@mail.ru
[2] Belarusian State Economic University, Minsk, Republic of Belarus
[3] Irkutsk State Agrarian University named after A.A. Ezhevsky, Irkutsk, Russian Federation
[4] Banzarov Buryat State University, Ulan-Ude, Russian Federation

Abstract. Digitalization of the economy is a major of priorities. Effective innovative development of regional and national economies requires all economic activities to be digitalized. In the current economic conditions, the issue of digital inequality between regions due to the low level of development of regional digital infrastructures is crucial. This has a negative impact on both on the innovative development of enterprises and on the standard of living and quality of life of every citizen in the country. The issue of digital inequality between the Russian regions requires both research and practical measures. Intensive implementation of information and communication technologies and equalization of the level of digital development can improve the efficiency of regional economies, the standard of living of people and positions of the Russian regions in business and social rankings. The article aims to identify features of digitalization of the regional economies, taking into account peculiarities of their innovative development. To solve the tasks set, an algorithm for evaluating the level of regional innovative development was developed, taking into account peculiarities of digitalization of regional economies. The algorithm is aimed at leveling inter-territorial differences in digitization of the Russian regions and boosting their innovation potential.

Keywords: Digitalization · Innovative development · Digital inequality of regions

1 Introduction

One of the main priorities of innovative development in Russia is digitalization of the national economy. The level of digitalization determines the country's competitiveness. To reach a higher level of economic development, Russia needs new scientific solutions and advanced developments. Those economic areas where the powerful technological potential accumulates should be developed using innovative technologies.

To update the technological system, it is necessary 1) to produce conditions for the competitiveness of the economy and to enter new markets, 2) to identify skills that allow

© The Author(s), under exclusive license to Springer Nature Switzerland AG 2022
A. Gibadullin (Ed.): DITEM 2021, LNNS 432, pp. 224–235, 2022.
https://doi.org/10.1007/978-3-030-97730-6_20

employees to take advantage of the digital economy, and 3) to determine the institutional framework for achieving maximum effect [7].

Digital technologies accelerate the pace of innovative development. Also it is necessary to take federal measures that can accelerate digitalization processes. These include reducing costs, investing in key business-related infrastructure facilities, reducing trade barriers, facilitating the entry of new companies into the market, strengthening antitrust authorities and encouraging competition between digital platforms [16, 17, 24].

The economy can be digitalized by automating all data processing processes and technologies. The main tools used to create an information society are presented in Fig. 1.

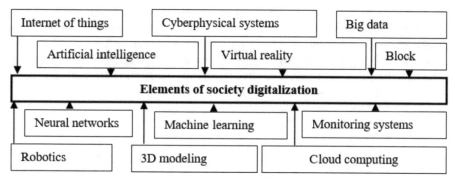

Fig. 1. The key elements of society digitalization (compiled by the authors).

There are several technologies which can contribute to the strategic development of national economies.

- Platform companies that are key links in the innovative economy. Platform companies help implement value propositions for the client based on open source solutions, machine learning, and cloud technologies with a required level of security;
- Industrial Industry 4.0 technology. It aimed at the greater cooperation of "cyberphysical systems" (CPS) into business operations;
- New logistics which should be based on standard infrastructure solutions such as Uber, car-sharing, drones, etc.;
- Digital money and new financial technologies aimed at creating a two-tier banking system;
- Technologies such as the neurocomputer interface, biotechnology, genetic engineering, etc. [8, 10, 13].

Equally important is the cultural environment. Its basis is federal, regional and local legal acts, standards, and rules.

The digital steps to mobility, sociality, big data and cloud computing are more than just technologies to be implemented. They involve the expansion, and in some cases the replacement of traditional operating models and processes with digital ones.

However, the digital divide is a major development issue for countries, regions and other territories. This problem is relevant for the Russian economy and economies of other developed and developing countries. This issue has been studied by a large number of researchers [1–23].

Thus, the article aims to determine the level of digitalization of Russia in the current economic conditions, to develop an algorithm for digitizing the regional economies taking into account the level of their innovative development and to eliminate digital inequalities.

2 Materials and Methods

The research methods used in the present study are data analysis and comparison. Using data from official statistics, the level of digitalisation and innovation development of both the our country and its individual territories has been analysed.

In order to bring the Russian economy to a new technological cycle of development - digitalization and high technology implementation, the sectoral structure of GDP should be based on information activities. The share of economic development expenditures as a percentage of GDP is presented in Fig. 2.

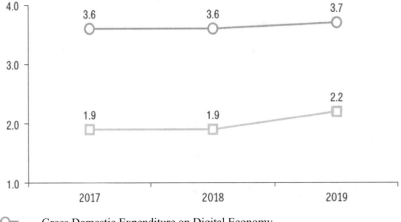

-○- Gross Domestic Expenditure on Digital Economy

-□- Internal expenditures on the creation, distribution and use of digital technologies and related products and services

Fig. 2. Expenditures on the development of the digital economy as a percentage of GDP [9].

The share of gross domestic expenditures on the digitalisation of the economy is about 3.5%. Since 2017, it has increased by 0.1%. The data indicate the need to create more favorable conditions for improving competition in the field of information and communication technologies (ICT). It is the ICT sector that will contribute to a significant inflow of targeted domestic and foreign investments.

There are several basic elements of the digital economy: online services, online commerce, industrial Internet, electronic banking and other payments, crowdfunding, etc. The development of information technology allows national economies to simplify the process of posting payments, to improve productivity, which, in turn, contributes to the development of online activities of companies reducing paperwork and production costs [2, 6, 12].

One of the main statistical indicators of the degree of ICT development is the number of Internet users and its dynamics (see Fig. 3).

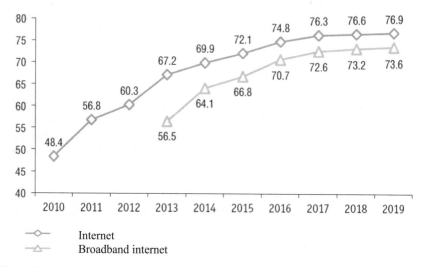

Fig. 3. Internet access in households (as a percentage of the total number of households) [9].

According to the 2019 survey, the level of Internet penetration in Russia has grown by almost 60 percent compared to 2010. At the same time, since 2017, this growth has slowed down, and averages about 0.3% a year.

By the total number of Internet users, Russia lags behind many developed countries (see Fig. 4).

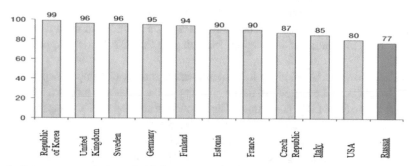

Fig. 4. Internet access in households by country: 2019 (as a percentage of the total number of households) [9].

For example, in the Republic of Korea, Great Britain, Sweden, and Germany, the Internet coverage is more than 95%. Figure 4 shows that Russia ranks eleventh. The solution to this problem is time-consuming and requires a developed information infrastructure. At the same time, to digitalize the country, it is necessary to develop information infrastructure in all Russian regions.

It is important to study features of digitalization in the Russian regions, taking into account their innovation potential. On the basis of our research, we have developed complex algorithm. This algorithm allows us to analyse the level of digitalisation of the regions, taking into account the specifics of their innovation development.

3 Results

To create a strong digital economy, it is necessary to encourage innovation activities. Innovation is considered to be an important source of business competitiveness.

The level of investment in information and communication technologies has a direct impact on the efficiency of innovation activities. The main performance indicators of the ICT sector are presented in Table 1.

Table 1. The main indicators of the ICT sector [compiled by the authors on the basis of statistical data 9].

Indicators	2015	2016	2017	2018	2019
Number of staff as a percentage of total staff	1.7	1.7	1.7	1.7	1.7
Gross value added as a percentage of GDP	2.8	2.8	2.9	2.8	2.8
Fixed capital investment as a percentage of the total volume of investment in fixed assets	3.1	3.1	3.0	3.4	3.9

Table 1 shows that the total number of employees in the various ICT companies does not exceed 2% of the total number of employees. At the same time, this indicator has remained stable over the past five years. The volume of investment in information and communication technologies does not exceed 4% of their total volume in fixed capital.

The main indicators of innovation, including in the ICT sector, are presented in Table 2.

Table 2 shows that in 2019 the level of innovative activity of companies decreased to the level of 2010. At the same time, the level of innovative activity of industrial enterprises was higher than that of ICT companies. The innovation activity of ICT companies tended to decrease. In the period under study, the decrease was more than 30%.

Equally important is the issue of digital inequality and different levels of innovation potential in the Russian regions. Digital inequality (or digital divide) means different opportunities to provide access to information and communication technologies.

The main indicators of society digitalization in the Federal Districts of the Russian Federation are presented in Fig. 5 and Fig. 6.

Table 2. The main indicators of innovative activity of Russian companies [compiled by the authors on the basis of statistical data 9].

Indicator	2010	2016	2017	2018	2019
Level of innovation activity, %	9.5	8.4	14.6	12.8	9.1
Industrial production	10.8	10.5	17.8	15.6	15.1
Telecommunications; computer software development, consulting services; information technology development	13.6	9.3	12.4	9.5	9.8

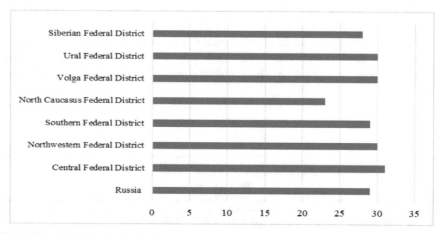

Fig. 5. The digitalisation index in the Federal Districts of the Russian Federation in 2019 [compiled by the authors on the basis of statistical data 9].

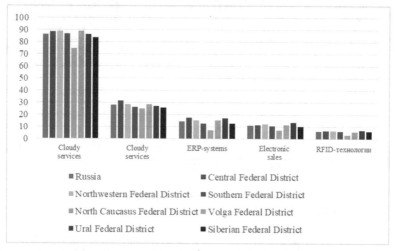

Fig. 6. The main indicators of society digitalization in the Federal Districts of the Russian Federation in 2019 [compiled by the authors on the basis of statistical data 9].

Table 3 shows that there are regional differences in the level of digitalization of both companies and people. The highest level of digitalization is observed in the Central Federal District, and the lowest one - in the North Caucasus Federal District. The Volga and Siberian Federal districts show high rates of digitalization.

The issue of digital inequality is characteristic of most developing digital economies. For example, the USA and Japan gain more than 1/3 of profits of the global ICT market, European countries - 1/3, and other countries - less than 1/3.

The digital divide leads to unequal development of the country's territories according to various indicators of social, economic, innovative and investment development. The level of digitalisation of a territory is connected to the level of its economic development: the higher the one, the higher the other. As a rule, a decrease in the level of economic development is associated with a decrease in digital potential.

Translated with www.DeepL.com/Translator (free version)Lack of developed information and communication infrastructure leads to a decrease in business activity. Moreover, the low level of digitalization has a negative impact on the quality of population's life.

The analysis identified causes of digital inequality between the Russian regions (see Fig. 7).

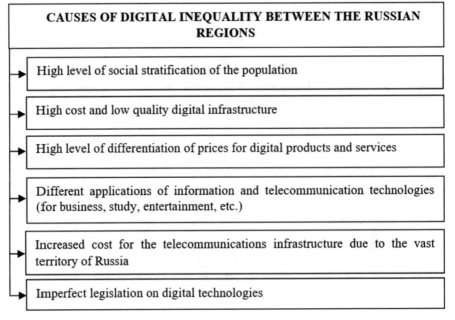

Fig. 7. Causes of the digital divide in the country's territories, using Russia as an example (compiled by the authors).

One of the main causes of digital inequality is social stratification of the population. Population incomes vary significantly. This affects the ability to buy and use digital technologies. In addition, the prices of digital products and services differ in different regions.

The Russian regions are characterized by different levels of investment and entrepreneurial activity in the ICT sector which has a direct impact on the financial opportunities for implementing digital technologies.

The vast territory of Russia also has a direct impact on the amount of costs required to create a telecommunications infrastructure. The lack of digital infrastructure facilities in remote areas creates a differentiation of digital products and services.

To increase the level of digitalization taking into account peculiarities of regional innovative development, it is necessary to develop an algorithm intended to assess the regional innovation potential and features of digitalization.

When assessing the level of innovation potential of the country's regions, we must examine their level of digitalisation. There are several basic indicators for assessing the innovation potential of territories and the level of its digitalization (Table 3).

Table 3. Indicators of digitalisation and innovation development of countries, regions, municipalities [compiled by the authors].

Potential	Indicators
Labor	Level of education of the population; number of researchers
Research	number of organizations performing research tasks, training graduate and doctoral students; number of applications and patents
Economic	Number of innovative enterprises; volume of investment; volume of production
Financial	Volume and cost structure for innovations
Infrastructure	Innovative infrastructure, number of technopolises, technoparks, information and analytical centers
Digitalization	Digital infrastructure, number of Internet users, Internet access in households, use of information and telecommunication technologies by companies, cost of digital economy development as a percentage of domestic regional product

The algorithm intended to assess the level of innovative development of the regions taking into account peculiarities of their digitalization is presented in Fig. 6 (see Fig. 6).

The method of structuring the indicator space as an ordered system of territorial clusters was used as a basis of this algorithm. In addition, the algorithm involves three stages.

The first stage involves the creation of a system of indicators used assess the level of digitalization of the region by the level of its innovative development. The correlation-regression analysis can be conducted to select the indicators that have the greatest impact on the processes under study. It helps identify the existing patterns of development and causal relationships between the economic indicators. It can also identify known and unknown dependencies and express them mathematically by constructing appropriate mathematical models used for explanation, prediction and control.

The second stage involves the assessment of the level of digitization and innovation potential of the region. For each indicator, average values are calculated. The indicators are compared with their calculated values. As a result, the regions can be classified by the level of their innovative development, taking into account peculiarities of digitalization of the regional economy (Fig. 8).

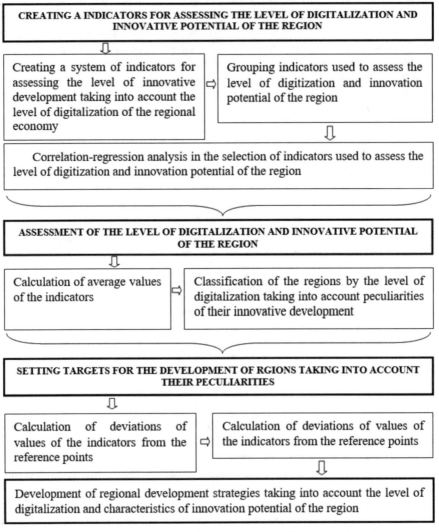

Fig. 8. The algorithm for assessing the level of innovative development of the region taking into account peculiarities of digitalization of the regional economy (compiled by the authors).

The third stage involves the identification of development benchmarks for all the indicators for each region.

The algorithm can be used to assess the level of digitization of regions taking into account peculiarities of their innovative development. This algorithm is aimed at leveling inter-territorial differences in the level of digitization and increasing the innovation potential of regions.

4 Discussion

The digital economy is a new economic area changing the existing ties and system models and methods of economic management. The current global digitalisation policy aims to create a global telecommunications infrastructure, a digitally enabled business environment, and a workforce capable of working in the new economic environment [11, 14, 18].

The analysis identified areas for increasing the level of digitalization of the regions, taking into account peculiarities of their innovative development (see Fig. 9).

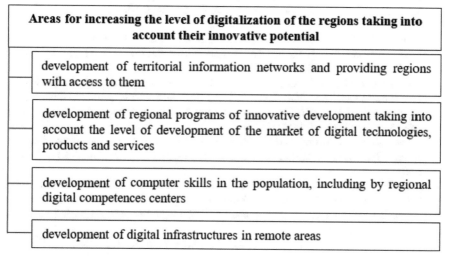

Areas for increasing the level of digitalization of the regions taking into account their innovative potential

development of territorial information networks and providing regions with access to them

development of regional programs of innovative development taking into account the level of development of the market of digital technologies, products and services

development of computer skills in the population, including by regional digital competences centers

development of digital infrastructures in remote areas

Fig. 9. Areas for increasing the level of digitalization of the regions taking into account their innovative potential (compiled by the authors).

A large number of foreign researchers have been dealing with the issue of digital inequality between the regions. This issue has been investigated at the regional, federal and international levels. At the same time, measures aimed to reduce the digital inequality between the regions and increase the level of their innovation potential can be an avenue for further research.

5 Conclusion

The qualitative growth of the economy requires technologies to accurately assess the current state of markets and industries, to predict their development trends and to ensure rapid response to changes in the domestic and global markets.

Federal and municipal authorities, scientific and educational communities and companies are addressing the digitalization of national and regional economies. An equally important issue is related to solving the problems of digital inequality of territories of different levels: countries, regions, municipalities, taking into account their innovation potential.

The process of digitalization affects all the existing and new markets, most of which will be networked. Russia is focusing on those markets that provide an opportunity to create technology industries that can ensure national security and high standards of living of people.

References

1. Antipin, D.A., Trufanova, S.V.: Project financing as a tool to enhance the role of commercial banks in the construction industry. IOP Conf. Ser. Earth Environ. Sci. **751**(1), 012130 (2021)
2. Antipina, O.V., Velm, M.V.: Characteristics of project management in the construction industry of the Russian Federation in modern economic conditions. IOP Conf. Ser. Earth Environ. Sci. **751**(1), 012072 (2021)
3. Avetisyan, K., Zhigalov, K., Gavrilova, Z., Salgiriev, A., Gaziev, V.: Improvement of the use of computer vision through the introduction of neural networks with clustering. J. Phys. Conf. Ser. **1582**(1), 012006 (2020)
4. Barykina, Y.N., Chernykh, A.G.: The leasing development tools in the construction industry of the Russian Federation. IOP Conf. Ser. Earth Environ. Sci. **751**(1), 012133 (2021)
5. Barykina, Y.: Analysis of information support for innovation development. IOP Conf. Ser. Mater. Sci. Eng. **667**(1), 012012 (2019)
6. Boikova, T., Zeverte-Rivza, S., Rivza, P., Rivza, B.: The determinants and effects of competitiveness. The role of digitalization in the European economies. Sustainability **13** (21), 11689 (2021)
7. Chetty, K., Aneja, U., Mishra, V., Gcora, N., Josie, J.: Bridging the digital divide in the G20: skills for the new age. Economics **12**(1) (2018)
8. Curran, D.: Risk, innovation, and democracy in the digital economy. Eur. J. Soc. Theory **21**(2), 207–226 (2018)
9. Digital Economy: 2021: a brief statistical digest. https://issek.hse.ru/news/420475066.html. Accessed 15 Nov 2021
10. Erkul, A.K.: Digital nomadism as a new concept at the intersection of GIG work, digitalization and leisure. Parad. Shifts Commun. World **170**, 49–59 (2021)
11. Khan, I.S., Ahmad, M.O., Majava, J.: Industry 4.0 and sustainable development: a systematic mapping of triple bottom line. Circular economy and sustainable business models perspectives. J. Clean. Prod. **297**, 126655 (2021)
12. Kobylkin, D, et al.: Network landscape representation: ecosystem services context. IOP Conf. Ser. Earth Environ. Sci. **751**, 012010 (2021)
13. Kuklina, M.V., Kuklina, V.V., Bogdanov, V.N., Starkov, R.F.: Extension and risks of development of social and technical networks in tourism. In: Proceeding of the International Conference on Research Paradigms Transformation in Social Sciences, RPTSS, pp. 651–658 (2018)
14. Kunkel, S., Tyfield, D.: Digitalisation, sustainable industrialisation and digital rebound – asking the right questions for a strategic research agenda. Energy Res. Soc. Sci. **82** (2021)
15. Maresova, P., et al.: Consequences of industry 4.0 in business and economics. Economies **6**(3), 46 (2018)

16. Nechaev, A.S., Antipin, D.A.: Mechanism for assessing the efficiency of financing the enterprise innovative activities. Actual Prob. Econ. **154**(4), 233–237 (2014)
17. Nechaev, A., Antipina, O.: Taxation in Russia: analysis and trends. Econ. Ann.-XXI **1–2**(1), 73–77 (2014)
18. Okrepilov, V.V., Gridasov, A.G., Chudinovskikh, I.V.: Standardization and metrology in the period of digitalization of the economy. J. Phys.: Conf. Ser. **1889**(3) (2021)
19. Pasqualino, R., Demartini, M., Bagheri, F.: Digital transformation and sustainable oriented innovation: a system transition model for socio-economic scenario analysis. Sustainability **13**(21), 11564 (2021)
20. Tikhomirov, A.A., et al.: Converting network–unlike data into complex networks: problems and prospective. J. Phys.: Conf. Ser. **1661** (2020). (International Conference on Information Technology in Business and Industry 2020)
21. Zakharov, S.V., Troshina, A.O., Lobova, A.U.: The method of selection of innovative solutions based on an assessment of the efficiency reserves. In: Proceedings of the 2017 International Conference IT and QM and IS 2017, pp. 601–602 (2017)
22. Zakharov, S., Shaukalova, A.: Methodological aspects of optimization of small enterprises in modern conditions of the Russian economy. IOP Conf. Ser. Mater. Sci. Eng. **667**(1) (2019)
23. Zhigalov, K., Gavrilova, Z.L., Daudov, I.M.: Creation of optimal cross-country route using GIS. J. Phys.: Conf. Ser. **1582**(1) (2020)
24. Nechaev, A., Prokopyeva, A.: Identification and management of the enterprises innovative activity risks. Econ. Ann. XXI **5–6**, 72–77 (2014)

Methodological Assessment of Indicators of Digital Transformation of the Russian Economy

A. F. Shupletsov[1]($\boxtimes$), Xu Zhanfeng[2], and Gou Lingyu[2]

[1] Baikal State University, Irkutsk, Russia
shupletsovaf@mail.ru
[2] Irkutsk National Research Technical University, Irkutsk, Russia

Abstract. The article deals with the conceptual issues of methodological evaluation of indicators of digital transformation of the Russian economy related to new available technologies: big data analytics and machine learning, artificial intelligence, robotics, augmented reality and the Internet of Things. An analysis of the applicability of digital solutions was conducted, the investment budget of companies was disclosed, and the expected payback period of investments in digitalization in the Russian Federation and foreign countries was assumed. This study identifies the problems faced by Russian companies that implement innovative and digital technologies. A model of the factors that affect the speed of digital transformation has been drawn up, and the authors reflect a scheme for assessing the effects and results of digital transformation. A methodology for calculating the digital transformation index to achieve effective results in assessing transformation indicators through the basic coefficients defined by the program "Digital Russia Indexes" was proposed.

Keywords: Digital transformation · Economy · Digital Russia Indexes

1 Introduction

The boundaries between the physical and virtual worlds are becoming more and more blurred. Technologies such as portable electronic devices, Internet objects, artificial intelligence and machine learning are changing companies and industries. The amount of data is growing faster than ever.

Companies expect these data to be used to create added value and provide competitive advantage. However, employees want to have simple and effective tools for working with data to achieve specific business goals. IT departments need to provide transparent access to data, while adhering to increasingly stringent security standards. New technologies and trends can be applied by different companies in different industries [16]. The competitive environment is changing, more and more companies need a stimulus from traditional competitors (companies that have long been working in the market) and beginners (startups and large technology companies that are developing new markets) [14]. In many cases, there are no economic barriers that would prevent small

© The Author(s), under exclusive license to Springer Nature Switzerland AG 2022
A. Gibadullin (Ed.): DITEM 2021, LNNS 432, pp. 236–246, 2022.
https://doi.org/10.1007/978-3-030-97730-6_21

businesses from entering the market - it is possible to acquire the same IT capacity that was previously available only to large companies that have invested in their technological development. In addition, such companies often have more opportunities for innovation and greater potential for a real revolution not only among competitors but throughout the industry. These innovators enjoy benefits of the digital transformation and are not hampered by outdated technologies and production facilities. Companies that do not use this opportunity are not able to survive. The McKinsey study found that in 1958, the average life expectancy of Standard & Poor's 500 companies was 61 years. However, it has been reduced to 18 years. According to McKinsey, by 2027, 75% of the companies will disappear or will be forced to unite with more successful competitors [2, 11].

These changes create new challenges for IT departments. They are no longer cost centers funded by enterprises which provide information support services. Information technologies have become an integral part of business processes, making the company faster, more flexible and competitive.

Russian and foreign companies are looking for a right path, as they gain experience through negotiations and mistakes. If foreign experience has already been taken into account in a number of studies, large-scale studies of Russian digital transformation management practices have not been conducted [19].

2   Materials and Methods

Digital transformation can improve business productivity and value of business—customer interactions, internal operational processes, or business models. The article presents statistical data on the implementation of digital technologies by Russian and EU companies.

3   Results

To implement a digital transformation program, managers should have a clear understanding of their digital strategies. Companies provide a wide range of services, which can be used in the integration of human resources, processes and technologies. According to the global KPMG survey, 70% of foreign executives believe that their organizations have a well-designed transformation agenda. 63% of executives of the largest Russian companies believe that their companies have digital transformation programs. Over the years, it has been difficult to find companies that have already tested or implemented digital technologies such as RPA and forecasting tools. Since 2019, the number of such companies has doubled, and the Russian market is ready for digital transformation processes. In Russia, there are companies that have clear digital transformation plans, and those that do not have a specific plan for business digitalization. Most companies in the telecommunications, financial and metallurgical sectors believe that they have digital transformation programs. On the other hand, companies in the transport and logistics sectors have no specific plans to digitize their businesses.

Many companies have been implementing digitization initiatives individually rather than as part of a comprehensive plan [1]. Projects with different goals have been implemented simultaneously, but they rarely coincide [20]. The situation can be improved by creating a structure in which project management will be performed within individual departments. Russian companies claim the existence of transformation programs, which can often be seen as an actual package of pilot projects that need to be implemented. In fact, most of them do not have medium- or long-term action plans [5, 6]. Instead, companies have short-term pilot projects. This approach allows us to assess the practical applicability of digital solutions, but often causes a shift to secondary tasks, a dispersion of resources and a suboptimal transformation of business processes.

In comparison with foreign EC countries, the Russian Federation lags far behind in the development of digital technologies. The difference is 10% [7].

The investment budget that Russian companies are ready to use for implementing digital technologies is presented in Fig. 1.

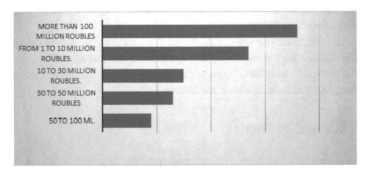

Fig. 1. Investment budget of companies that have disclosed their investment data.

IT integrators and transport companies are the least willing to invest. For example, 66% of the IT companies are willing to allocate less than 30 million rubles a year. Not all companies know the total volume of investments into the digital transformation programs, as most of the pilot projects have already been implemented. Companies that have developed comprehensive 3–4-year programs say that the funding of approved projects may exceed one billion rubles. Companies, that developed digital transformation programs four years ago, have already invested more than one billion rubles [9].

According to the KPMG global survey, 42% of foreign corporate managers expect a return on investment in less than 2 years, while 30% of companies expect a return within 12 months (see Fig. 2).

Russian companies are more conservative than EU ones: only 51% expect a return on investment in less than 2 years. Managers of 45% of Russian companies prefer technologies that have already been used in the market. More than 45% of companies are willing to invest. 30% of global companies have a one-year horizon for digitization programs. For comparison, in the Russian Federation there are twice as many companies with such expectations than in the world, only 15% of Russian companies expect profits within 12 months. The expected average return on investment in digital technologies is two years. The more money companies invest in digital technology, the faster they will get positive results [18].

The main obstacles to digitization are insufficient maturity of current processes, low-level automation, insufficient competence and IT illiteracy of employees [12].

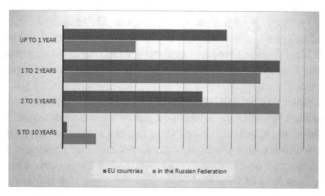

Fig. 2. The expected digitalization investment payback period in the Russian Federation and foreign countries.

Industrial enterprises indicate the insufficient level of development of automated process control systems (ACS) and production process control (MES) 2, which is also an important factor in the development of forecasting tools. Most market participants are still at the data collection stage. The executives argue that these patented production management systems are often installed by enterprises where the data is stored in their own unique or outdated formats, or encrypted, which makes it very difficult to obtain and accumulate information from them. In the future, master data management will be crucial for the subsequent digitization of processes [13].

Companies are aware of the need to improve the maturity and optimize business processes. To achieve positive results, the end-to-end business processes should be taken into account [15].

Figure 3 shows barriers encountered by Russian companies implementing innovative technologies.

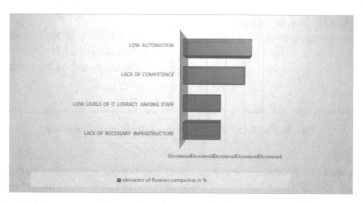

Fig. 3. Barriers encountered by Russian companies implementing innovative technologies.

As can be seen, one the main obstacles is a low level of automation of current business processes and skills and competencies [4].

According to the Organization for Economic Development and Cooperation (OECD), two key factors affect the speed of digital transformation. The first factor is internal capabilities (human resource management and proper allocation of resources).

The second factor is a level of competition in the industry, availability of technology and capital, and legislation [3].

There is a direct link between the level of development of human capital and effectiveness of implementation of digital technologies. Successful implementation of digital initiatives requires strong leadership skills and modern management methods. Effective integration of technologies and business processes requires employees to have basic IT skills, maintain and develop these skills and match the nature of work with their competencies. Managers of large Russian companies argue that skills are an important indicator [10].

Large-scale restructuring programs are aimed at improving the digital literacy of employees. This is especially important for non-IT employees, namely business departments. In general, estimates of the Russian companies correlate with the data of international studies, in particular with the OECD analysis conducted in 2018 [17].

Figure 4 shows that there is a direct link between the development of human capital and the effectiveness of implementation of digital technologies.

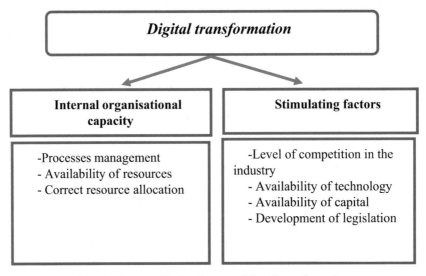

Fig. 4. Factors affecting the rate of digital transformation

Participants in the Russian survey mentioned information security as a threat to digitalization. 56% of Russian executives are concerned about information security issues; according to 68% of EU company executives, this threat ranks first. Despite the fact that 46% and 64% of Russian corporate managers have concerns about the risk of unemployment, the EU executives believe that digital technologies can create more jobs in five years.

According to Russian companies, the biggest threat to digitalization is information security managers.

According to the KPMG global survey, 95% of Western corporate executives believe that a technological breakthrough offers more opportunities than threats.

The executives believe that the digital transformation can increase the number of jobs (42% of EU company executives vs 58% of Russian company executives).

Only a few companies can achieve new, higher cost indicators. Companies are more likely to implement IT pilot projects that are insufficiently coordinated with each other. This is a way of digitization without a clear vision of the future. Pilot projects inspire unwarranted optimism, while leaders believe that costly initiatives are lagging behind [3].

According to the KPMG's global study, the most effective approach to digital transformation is to use a "business objectives" methodology, which involves the determination of visible outcomes and sources and selection of a specific implementation technology. This is different from the traditional approach, according to which assessment of project results is preceded by project implementation. The most effective approach is the business task method. When a company implements new technologies, it first determines what results should be achieved and sources should be used, and then selects a specific technology [17] (Fig. 5, 6 and 7).

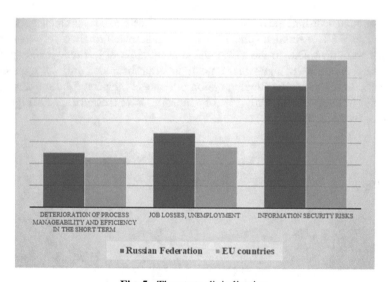

Fig. 5. Threats to digitalization.

Figure 6 shows that 52% of foreign executives and 48% of Russian ones believe that the development of digital technologies opens up more opportunities than threats.

The main approaches can be borrowed from the leading Russian companies that have already implemented transformation programs.

The focus is on costs and results rather than on the technology itself. When choosing projects, market leaders with limited resources prioritize processes and areas that

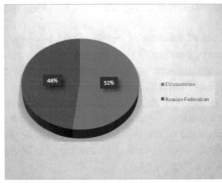 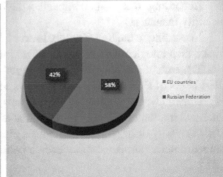

Fig. 6. Opportunities for the development of digital technologies

Fig. 7. The increasing number of jobs due to the implementation of digital technologies

can increase the value of business [8]. The path to successful transformation depends on a balanced strategic approach to the implementation of digital initiatives aimed at increasing the value of business.

Most Russian companies have failed to develop a decision-making process for financing and allocating their budgets for implementing specific digital transformation projects. At present, such decisions are made on an individual basis, and senior managers analyze the initiatives. 36% of Russian companies fund digital transformation projects from the development budget. Due to the traditional concentration of technological experience in IT departments, 25% of companies fund digital transformation projects together with IT projects. Another popular funding option is a budget intended for digitizing departments (20% of respondents). Only 13% of companies have created a special digitization committee (see Fig. 8).

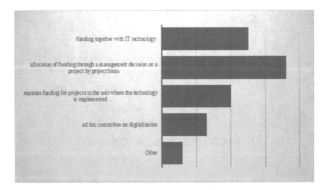

Fig. 8. Funding of digitization projects in Russian companies.

Of Russian 36% companies fund their digitization projects through the decision-making process at the top management level. The share of funding IT technologies is 25%

In general, the digital transformation functions are distributed among company departments [5, 6].

A model in which specific employees bear responsibility for implementing such programs has not yet been developed. In 44% of Russian companies, top managers are responsible for implementing digital initiatives (Fig. 9).

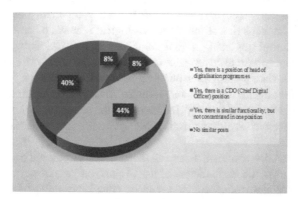

Fig. 9. The position of a digital officer in all Russian companies

Only 16% of Russian companies have a position of CDO (Chief Digital Officer) or a manager of digitization programs. The CDO is a position that has recently appeared in Russia. Only 8% of companies (metallurgical and telecommunications companies) have CDOs. There are positions with similar duties in financial and IT companies.

Most Russian companies have not yet created a center of competence in digital technologies (only 34%). Only 20% have competence centers with similar functionality. 14% of corporate managers believe that they have departments with similar functions.

If the digitization process is not realized by a specific department, the managers should create a mechanism for the exchange of knowledge and information within the company. For companies that have been testing digital solutions at the block level, it is important that participants in such projects are synchronized at the strategic level [12].

4   Discussion

The digital transformation of the country's economy, which covers all sectors and industries, poses new challenges to the industrial enterprises of the Russian Federation, which must provide an effective solution to three managerial tasks. Let us consider the information scheme of the digital transformation in Fig. 10.

Thus, according to the scheme shown in Fig. 10, we can identify the following key criteria for the effectiveness of digital transformation:

– Strategic goal-setting of digital transformation in industries, which is formed within the framework of the state system of strategic planning and includes the justification of sectoral technological priorities of digital transformation, rational levels of depth and scale of digital transformation, as well as an overall assessment of the costs and benefits of digital transformation and integral estimates of its effectiveness in the industries, the economy;

– The implementation of public policy during the digital transformation through a system of effective administrative;

– The implementation of government policy during the digital transformation of FEC industries through a system of effective administrative, regulatory, organizational, and economic mechanisms that stimulate technological renewal of production facilities, systems and complexes using digital solutions, as well as changes in the market environment and corporate governance (at least for companies with state participation, companies with state participation and occupying a key position);

– Monitoring of the results achieved, including indicators of the depth and scale of digital transformation and "cross-cutting" indicators of the results of digital transformation, allowing the assessment of key results achieved in different industries.

To achieve effective results in assessing the indicators of digital transformation, it is necessary to calculate the basic factors of the Digital Russia Index.

Indicators of digital transformation, can be calculated using the methodology of evaluation of basic factors. The present methodology calculates both quantitative and qualitative assessment of the index.

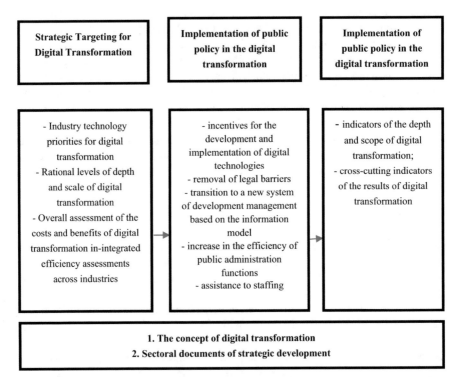

Fig. 10. Scheme for assessing the effects and results of digital transformation. Note: Designed by the authors.

Quantitative assessment of the index of digital transformation of the subjects of the Russian Federation is calculated using the integral criterion of the form:

$$K_0(t) = \alpha K_{np}(t) + \beta K_{ko}(t) + x K_{iktz}(t) + \delta K_{ii}(t) + \varepsilon K_{ib}(t) + \epsilon K_{oqc}(t) + \varphi K_{oqc} \quad (1)$$

$K_0(t)$ - the final value of the index, reflecting the effectiveness of public coverage in open sources of the level of digitalization, including the implementation in the subject of the Russian Federation of the program "Digital Economy of the Russian Federation Russian Federation" program and the solution of the main tasks of digitalization economy to create the conditions for the digital transformation of the infrastructure of all industries and social infrastructure based on the results of the analysis of information from open sources at the time of;

$K_{np}(t)$ - assessment of the level of public coverage in open sources of development in the subject of the Russian Federation normative regulation and administrative indicators digitalization, providing the processes of digitalization of the economy at the moment time;

$K_{ko}(t)$ - assessment of the level of public coverage in open sources of development in the subject of the Russian Federation the direction of specialized personnel and training programs providing the processes of digitalization of the economy at a time;

$K_{iktz}(t)$ - assessment of the level of public coverage in open sources development in the subject of the Russian Federation of the availability and formation of research competencies and of technological advances, providing the processes of digitalization of the economy at point in time;

$K_{ii}(t)$ - assessment of the level of public coverage in open sources of development in the subject of the Russian Federation of the direction of information infrastructure, providing the processes of digitalization of the economy at a point in time;

$K_{oqc}(t)$ - assessment of the level of public coverage in open sources financial and economic efficiency of the development of digitalization in the subject of the Russian Federation at point in time.

α, β, x, δ, ε, ϵ, φ - the evaluation coefficients of the level of public coverage in open sources of the development of digitalization areas: "regulatory regulatory framework", "human resources and education", "research competencies and technological reserves", "information infrastructure" and "information security", "financial and economic efficiency", "social efficiency", respectively, determined by the factor analysis and the method of expert evaluations, and satisfying the condition of rationing:

$$\alpha + \beta + x + \delta + \varepsilon + \epsilon + \varphi = 1 \quad (2)$$

Thus, the Digital Russia Index makes it possible to obtain an expert The Digital Russia Index makes it possible to obtain an expert evaluation of this parameter in the context of each constituent entity of the Russian Federation, the accuracy and reliability of which correspond to the accuracy and reliability of the information which corresponds to the accuracy and reliability of the information obtained from public sources.

5   Conclusion

The process of digital transformation is no longer something new for Russian companies and is crucial for their success: only a quarter (26%) of respondents believe that they have resources and competencies to implement digitalization plans. In 2019–2021, the most common areas of investment are cybersecurity, multi-cloud environments, flash memory technologies and Internet facilities.

Digital platforms are the most discussed topics. Economists and politicians, IT professionals and businessmen, researchers and teachers have been discussing their key role in transforming reality. It is time to develop the digital environment in all spheres. The focus should be on promising innovative technologies such as blockchain, artificial intelligence, big data analysis, etc. Unfortunately, attention is not paid to fundamental technologies.

There are more and more digital technologies. Each person has a large number of digital devices with functions which are not always known to consumers. Some electronic devices and computer programs are a black box.

However, it is evident that digital technologies will conquer new frontiers and people will not be able to avoid the widespread use of these technologies. They may be afraid of this process, but it is useless to resist.

The relatively rapid development and cheapness of digital devices means that the digital methods of recording and processing information will soon displace the analog ones. These technologies will affect the world around us. This confirms the idea that the road is not close and that this is the most interesting thing we have not seen.

References

1. Andreeva, E.S., Nechaev, A.S.: The mechanism of an innovative development of the industrial enterprise. World Appl. Sci. J. **27**(13 A), 21–23 (2013)
2. Antipina, O.V., Velm, M.V.: Characteristics of project management in the construction industry of the Russian Federation in modern economic conditions. IOP Conf. Ser. Earth Environ. Sci. **751**(1), 012072 (2021)
3. Cao, S., Nie, L., Sun, H., Sun, W., Taghizadeh-Hesary, F.: Digital finance, green technological innovation and energy-environmental performance: evidence from China's regional economies. J. Clean. Prod. **327**, 129458 (2021)
4. Chyzhevska, L., Voloschuk, L., Shatskova, L., Sokolenko, L.: Digitalization as a Vector of Information Systems Development and Accounting System Modernization Studia. Universitatis Vasile Goldis Arad, Economics Series (2021)
5. Erol, I., Peker, I., Ar, I.M., Turan, I, Searcy, C.: Towards a circular economy: investigating the critical success factors for a blockchain-based solar photovoltaic energy ecosystem. Turkey Energy Sustain. Dev. **65**, 130–143 (2021)
6. Grau-Sarabia, M., Fuster-Morell, M.: Gender approaches in the study of the digital economy: a systematic literature review. Hum. Soc. Sci. Commun. **8**(1), 201 (2021)
7. Huang, Y., Chen, Y., Tan, C.H.: Regulating new digital market and its effects on the incumbent market. Investigation of online peer-to-peer short-term rental. Inf. Manage. **58**(8), 103544 (2021)
8. Ilina, E., Tyapkina, M.: Enterprise investment attractiveness evaluation method on the base of qualimetry. J. Appl. Econ. Sci. **11**(2), 302–303 (2016)
9. Kunkel, S., Tyfield, D.: Digitalisation, sustainable industrialisation and digital rebound – asking the right questions for a strategic research agenda. Energy Res. Soc. Sci. **82**, 102295 (2021)

Creation of a Digital Model of Federal, Regional or Intermunicipal Public Roads

Victoria Shamraeva[1] , Vitaly Mironyuk[2], Evgeny Savinov[1] ,
Lyudmila Rudenko[3] , Dmitry Morkovkin[1(✉)] , and Elena Kolosova[4]

[1] Financial University Under the Government of the Russian Federation, 49,
Leningradsky Avenue, Moscow 125993, Russia
VVShamraeva@fa.ru
[2] Federal Autonomous Institution "Russian Road Research Institute", 2, Smolnaya Street,
Moscow 125493, Russia
[3] Moscow Witte University, Building 1, 12, 2-nd Kozhukhovskiy proezd, Moscow 115432,
Russian Federation
[4] Plekhanov Russian University of Economics, 36, Stremyanny Lane, Moscow 117997,
Russian Federation

Abstract. The abstract should summarize the contents of the paper in short terms, i.e. 150–250 words. Introduction: The object of the study is a public road network. The subject of the study is a digital public road model. The implementation of an expanded concept for a digital public road model creation and the ways of its development are currently regulated by the action plan of December 18, 2017 in the direction of "Formation of research competencies and technological foundations" of the program "Digital Economy of the Russian Federation" and the measures provided for by the federal project "System-wide measures for the development of the road facilities" of December 20, 2018 and are priority areas of digital transformation of the road facilities and transport complex of the Russian Federation. To analyze the measurement results of the pavement longitudinal evenness of the public road control sections obtained using mobile laser scanning systems and mobile diagnostic laboratories; to evaluate the accuracy of determining the spatial position (planned and altitude position) of objects (elements of arrangement, defects, etc.) from the point cloud of laser scanning and photo, video images recorded respectively by mobile laser scanning systems and mobile road laboratories and the accuracy of determining the parameters of the public road geometric elements (section length, pavement width, pavement area; to analyze the measurement results of the pavement longitudinal evenness of the public road control sections obtained using mobile laser scanning systems and mobile diagnostic laboratories; to develop a unified system of views, approaches, definitions, etc. to form a methodological basis for creating a digital public road model as the basis for carrying out measures for the digital transformation of the road facilities and the transport complex of the Russian Federation. Identification of roadside infrastructure objects, arrangement elements; coating defect determination; geo-radar scanning profile analysis; mobile road laboratory use. The main directions are proposed for the development of an expanded concept for creating a digital public road model, as well as its use as an information basis for creating, storing and using data on public roads, taking into account the development planning of

© The Author(s), under exclusive license to Springer Nature Switzerland AG 2022
A. Gibadullin (Ed.): DITEM 2021, LNNS 432, pp. 247–255, 2022.
https://doi.org/10.1007/978-3-030-97730-6_22

the road transport network, design, construction and operation of the public road network.

Keywords: Digital public road model · Diagnostics of public roads · Infra-BIM · Laser scanning

1 Introduction

Within the framework of the measure implementation provided for by the federal project "System-wide measures for the development of road facilities" work on the creation of a digital public road model (hereinafter referred to as the digital public road model) (sections of public roads) of federal, regional or intermunicipal significance is relevant and should be completed by the end of 2024. In [1], the concept of creating a digital public road model (hereinafter referred to as the Concept) is laid down, which sets out general methodological provisions, approaches, definitions for forming a methodological basis for creating the digital public road model as the basis for carrying out measures for the digital transformation of the road facilities and the transport complex of the Russian Federation. In addition to the general methodological provisions, the Concept also defined the directions of using the digital public road model as an information basis for the creation, storage and use of data on public roads, taking into account the development planning of the road transport network, design, construction and operation of the public road network.

The development of the digital public road model corresponds to the measure plan of December 18, 2017 in the direction of "Formation of research competencies and technological reserves" of the program "Digital Economy of the Russian Federation" (Measure 03.00.000.001.18.11 - Creation of a software package for instrumental monitoring of the diagnostic public road condition, designed to collect, accumulate, store, consolidate, analyze and interactive visual representation of data on the condition of public roads received from various measuring and control equipment, management bodies and the population. The article considers the expanded concept of creating a digital public road model and defines the directions of using digital public road model as an information basis for creating, storing and using data on public roads, taking into account the development planning of the public road network, design, construction and operation of the public road network. The development of digital public road model, InfraBIM for the purposes of managing the business processes of the public road construction and operation [2] and their regulatory framework is extremely necessary and should contribute to the successful implementation of information modeling in Russia [3].

2 Subject, Tasks and Methods

Work on the creation of the digital federal, regional or intermunicipal public road model is carried out in two stages:

- Stage I (industrial testing) - it is planned to take measures to create a digital federal public road model (sections of public roads) in the amount of 20 thousand km. Deadline: January 1, 2020 - December 31, 2021.
- Stage II – (full-scale work) - carrying out measures to create a digital federal, regional or intermunicipal public road model. Due date: January 1, 2022 - December 2, 2024

In order to develop a unified system, views and approaches when creating a digital public road model, it is necessary to clearly define its composition and the requirements for equipping mobile road laboratories for collecting spatial data, data formats and mechanisms for uploading results to the control system for the formation and use of road funds at all levels.

The composition of the digital public road model and the requirements for equipment and mobile road laboratories should be justified by comparative tests of mobile laser scanning systems and mobile road laboratories. Such tests were carried out in two stages. The main tasks of the tests:

- To evaluate the accuracy of determining the spatial position (planned and altitude position) of objects (elements of arrangement, defects, etc.) from a point cloud of laser scanning and photo, video images recorded respectively by mobile laser scanning systems and mobile road laboratories and the accuracy of determining the parameters of geometric elements of the public road (length of the section, width of the pavement, area of the pavement [4] (tasks of the I stage);
- Analyze the results of measurements of the pavement longitudinal evenness of public road control section obtained using mobile laser scanning systems and mobile diagnostic laboratories and evaluate the accuracy of determining the IRI evenness index from the point cloud recorded by mobile laser scanning systems and from the results of measurements of sections by diagnostic laboratories [5] (tasks of stage II);

Thus, based on the results of comparative tests, it is possible to determine the minimum data set and form a data structure for creating a digital public road model (Fig. 1).

The point cloud of laser scanning results makes it possible to determine the geometric dimensions of the public road elements in the desk conditions, to obtain relief surfaces and to vectorize the public road elements and artificial structures, elements of arrangement and roadside service facilities [6, 7]. At the same time, these data are amenable to automated processing and analysis. And since each point of the cloud has spatial coordinates, all the information generated on the point clouds has a geographical and altitude reference.

3   Method Development

To create a digital public road model, it is necessary to identify objects of roadside infrastructure, elements of the arrangement and identify pavement defects, using geo-linked panoramic images. Information about the subsurface state will be formed as an analysis result of georadar scanning profiles [8]. Thus, the specified data set is the minimum necessary for carrying out desk work and subsequent problem solving, including

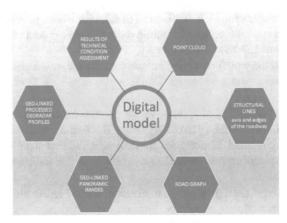

Fig. 1. Digital model data structure.

using various process automation tools. Further, as part of the desk work using laser scanning cloud data, structural lines should be constructed, that is, the axial and edges of the pavement and the graph. Structural lines are the initial data for design work for various purposes. The main purpose of the road graph is to ensure the connectivity and comparability of data on the road obtained in different time periods. As part of the work on the creation and refinement of the road graph elements to ensure the possibility of step-by-step filling with data and information of the public road digital model, six levels of its elaboration (reliability of information) are provided. At the lowest, sixth level, there is no geo-linking of materials, but the fact of the information availability itself has been confirmed. The most accurate is the first level, which contains information obtained as a result of geodetic measurements. The results of diagnostics and certification of the data obtained with sub-meter accuracy [9, 10] correspond to the third level. As part of the implementation of measures for the creation of the digital public road model, the information collected from the results of road laboratories, the trajectory of which is calculated in post-processing mode or using Precise Point Positioning technology, will correspond to the second level.

When forming the digital public road model, including when forming road graphs, it is assumed that the maximum use of already available data in digital form at the federal and regional levels, data from diagnostic results and data obtained as part of the implementation of measures for the national project "Safe and high-quality public roads" is assumed. To ensure the uniformity of measurements, all data is collected with reference to the coordinate system WGS-84 or GSK-2011.

Having analyzed the public road pavement by networks of reference stations, it was found that part of the public road network is not provided with the network coverage of such stations. If in the central part of Russia coverage in some areas reaches 90–100%, then beyond the Urals and in the Far East there are regions where only 10–20% of public roads are provided with reference stations [11]. Taking this fact into account, it is possible to determine the technology of field work production. As a basis, networks of base stations are used, which in turn are linked to the points of fundamental astronomical and geodetic network and IGS. In the absence of coverage, temporary base stations are installed at the work sites, the binding of which is carried out by static observations. Taking into account the completed binding to the points of fundamental astronomical and geodetic network and IGS, the coordinates and heights of temporary base stations can be determined in PPP mode. This technology of work makes it possible to combine data obtained in different time periods, in addition, it becomes possible to supplement the digital model with any other data that has an appropriate geographical reference. This ensures the accumulation of various data, performing their joint analysis, as well as using them to solve various tasks.

Mobile road laboratories are used to collect spatial data, which collect interconnected data by synchronizing all measuring devices with a single inertial navigation module. The use of such laboratories reduces inefficient runs and ensures the linking of various data at the physical level.

As part of the road laboratory (Fig. 2), there are: a mobile laser scanning system (1) with a panoramic camera that provides the formation of a panoramic image and three additional cameras, a multi-channel georadar (2), a laser recorder of longitudinal evenness (3). All measuring devices are synchronized with a single inertial navigation module of the mobile laser scanning system, and an odometer is used to record the distance traveled (4). The road laboratory is equipped with a place for the operator to work, equipped with an on-board computer for controlling measuring devices.

Separately, it is worth noting the need to use a laser scanning system with two scanners located at an angle to the direction of movement [7, 8]. Taking into account the length of the public road network, the most important requirement for a road laboratory is to obtain the maximum amount of complete data for the minimum number of passes. Figure 3 shows an example when a passing vehicle creates an obstacle for laser beams.

When using a laser scanning system with a single scanner located perpendicular to the movement direction (Fig. 3, a), a shadow zone is formed not only on the carriageway directly under the vehicle, but also outside the carriageway. The use of two scanners (Fig. 3, b) minimizes the size of the shadow zone behind moving vehicles, which increases the informativeness and integrity of the point cloud. Each road laboratory has a set of geodetic equipment for carrying out survey work, control measurements and providing the possibility of additional spatial data binding in local coordinate systems.

Branches of the FAI "ROSDORNII" and the Financial University under the Government of the Russian Federation will be involved in the work on the public road digital model creation, which will ensure uniform performance of work throughout the Russian Federation, as well as timely collection and updating of data, if necessary.

The collected data will be transferred to the head office of the FAI "ROSDORNII" for processing and forming a digital model. The generated digital model will be uploaded to the Road fund control system. The technology of spatial data collection at the stages of field and desk work was tested and refined during the creation of the public road digital model (road sections) at pilot facilities [10, 11].

For the organization and execution of work, a department for the collection and processing of spatial data has been specially formed, which provides, among other things, coordination and control of the activities of the units involved in the spatial data collection. As part of the measure implementation to create the public road digital model, mechanisms are also being worked out and possible sources of updating information are

Fig. 2. Road laboratory for spatial data collection: 1 - mobile laser scanning system, 2 - georadar, 3 - laser recorder of longitudinal evenness, 4 - odometer.

being determined, such sources should be the results of engineering surveys, executive surveys, the results of diagnostics of sections of public roads, as well as the results of special surveys. Timely updating of the data of the digital model in compliance with the principle from general to private will ensure the gradual filling of data and the integrity of information.

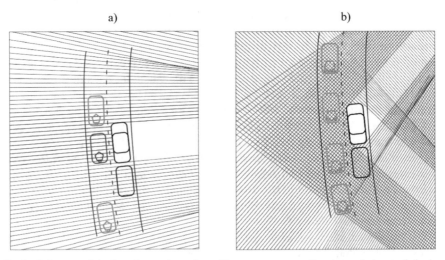

Fig. 3. Influence of the location and number of laser scanners on the size and shape of shadow areas.

Work on the spatial data collection on a certain route begins with the analysis of the network coverage of reference stations in the area of work. If the distance from the nearest reference station does not exceed 70 km, then the calculation can be performed directly from it, otherwise temporary base stations are installed at the work site and their coordinate and altitude reference is carried out.

Immediately before the start of spatial data collection, the attachments of the road laboratory are mounted and initialization is performed. Data is collected at a speed of 80 km/h. During the movement, the operator of the road laboratory monitors the operation of measuring devices.

The received data is recorded on SSD disks and sent to the head office of the FAI "ROSDORNII" for processing and creating a digital model. Desk work begins with the mutual equalization of base stations with each other. This is necessary in order to eliminate possible errors in the source coordinates, to equalize the base stations of various operators, and also to link temporary base stations. Then, the exact trajectory of the road laboratory is calculated from the equalized base stations in specialized software. The resulting trajectory is used to form point clouds (Fig. 4) and snap panoramic images. Structural lines are drawn using point clouds. Point clouds, panoramic images and rendered structural lines are further used to form a road graph. The previously obtained exact trajectory of movement is also used to link georadar profiles and diagnostic results. After that, the digital model is formed and the data is loaded into the Road fund control system.

Fig. 4. An example of a point cloud obtained as a result of mobile laser scanning.

The main directions in the implementation of which, taking into account the refinement and addition of missing data and information, the public road digital model can be used including:

– The formation of departmental navigation information resources for state needs and state information systems,

- Carrying out measures to automate processes in the road sector and ensuring management decisions for carrying out repair activities,
- The introduction of information modeling in the road facilities of the Russian Federation,
- Carrying out activities for planning the development of the road transport network (development of integrated transport schemes, work within the framework of territorial planning schemes, planning projects, work on modeling traffic flows, etc.)
- Ensuring the development of infrastructure for the movement of unmanned vehicles, as well as for use as:
- Initial data for initiating and carrying out design work,
- The basis for increasing the level of work automation in terms of public road certification, the project development for the transportation of bulky and heavy loads, the development of integrated traffic management schemes, traffic management projects, etc.

4 Conclusion

The proposed expanded concept of creating a general-use public road digital model of federal, regional or intermunicipal significance will become the basis for carrying out measures for the digital transformation of the road economy and transport complex and will increase the automation of construction and operation processes, traffic flow management and road safety. The main directions of the use of the public road digital model as an information basis for the creation, storage and use of data on highways are determined, taking into account the planning of the development of the road transport network, design, construction and operation of the public road network. The advantage of the created the public road digital model is obvious and cannot be underestimated for different categories of users [12].

Thus, for citizens and business representatives, the expected socio-economic effect of the creation of the public road digital model will be an increase in quality, accessibility and convenience, as well as a reduction in the delivery of public services and municipal services in digital form. This will be achieved by having a single source of data on public roads. For public sector bodies and organizations, the effect will be to increase the reliability of the collected data and the breadth of their coverage, increase the accuracy of planning and forecasting, speed and quality of management decisions, increase opportunities for the development of the digital economy through the creation of a digital data bank [13–16].

References

1. The concept of creating a digital model of public roads (road sections) of federal, regional or intermunicipal significance, an order providing for the use of modern diagnostic methods and mechanisms for uploading the results of such diagnostics to the Road fund control system. Moscow, 25 (2020)
2. Shamraeva, V., Savinov, E.: Infra-BIM for business processes' management in road construction and operation. Architect. Eng. **6**(3), 19–28 (2021)

3. Znobishchev, S., Shamraeva, V.: Practical use of BIM modeling for road infrastructure facilities. Architect. Eng. **4**(3), 49–54 (2019)
4. The length of public roads in the subjects of the Russian Federation for 2019. https://rosstat. gov.ru/storage/mediabank/iNEf3mAv/t2-2.xls. Accessed 14 Oct 2021
5. Mironyuk, V. P., Kuznetsov, A. O.: Comparative tests of mobile laser scanning systems and mobile road laboratories. Verkhnebakansky highway, Krasnodar, 106 (2020)
6. Mironyuk, V.P., Lushnikov, P.A.: Comparative tests of mobile laser scanning systems and mobile diagnostic laboratories. Research Center of NICIAMT FSUE NAMI, (second stage), Moscow, 44 (2019)
7. Kuznetsov, A.O.: Modern mobile laser scanning systems and features of their application on public roads. Roads Bridges **42**, 56–76 (2020)
8. Eremin, R.A.: Application of computer vision for interpretation of georadar data. Mir dorog World Roads **131**, 86–89 (2020)
9. ODM 218.4.039–2018: Recommendations for the diagnosis and assessment of the technical condition of highways. https://odm.ru. Accessed 15 Oct 2021
10. Report on the performance of works under. Moscow, 33 (2019)
11. Report on the performance of works under. Moscow, 36 (2019)
12. Shamraeva, V.V.: Predicting the effectiveness of implementation of financial models for toll collection on the road. Transp. Bus. Russ. **1**, 37–41 (2020)
13. Shamraeva, V.V.: Analysis of business processes of construction and operation of public roads on a toll basis using bim tools. In: Information Technologies and Mathematical Modelling: 19th International Conference, ITMM 2020, named after A.F. Terpugov, Tomsk, Russia, pp. 429–440 (2021)
14. Shamraeva, V.: BIM - modeling of transport infrastructure objects. In: International Conference E-Business Technologies EBT 2021, Belgrade, Serbia, pp. 29–30 (2021)
15. Gibadullin, A.A., Morkovkin, D.E., Hutarava, I.I., Stroev, P.V., Pivovarova, O.V.: Analysis and digital transformation of the transport sector of the Eurasian Economic Union. IOP. Conf. Ser. Mater. Sci. Eng. **918**, 012231 (2020)
16. An, J., Mikhaylov, A., Jung, S.-U.: A linear programming approach for robust network revenue management in the airline industry. J. Air. Transp. Manage. **91**, 101979 (2021)

Applied Methodology for the Formation of Predictive Financial Statements of an Industrial Enterprise, Taking into Account the Influence of the COVID-19 Factor

Ekaterina Kharitonova[1]([✉]), Natalia Kharitonova[1], and Ilia Litvinov[2]

[1] Financial University Under the Government of the Russian Federation, 49, Leningradsky Pr., Moscow 125167, Russia
eharitonova@fa.ru

[2] Limited Liability Company "Interkos-IV" (PJSC Magnitogorsk Iron and Steel Works and Subsidiaries), Office 334, House 122, Lit. A, Territory of Izhora Plant, Kolpino, St. Petersburg 196650, Russia

Abstract. The article describes an applied methodology for the formation of predictive financial statements of an industrial enterprise, taking into account the influence of the COVID-19 factor. Based on the data of the consolidated interim financial statements of PJSC Magnitogorsk Iron & Steel Works and Subsidiaries for 9 months ended September 30, 2021 and 2020, the authors performed forecast calculations for 9 months of 2022, taking into account adjustments to the actual reporting due to the influence of the COVID-19 factor in 2020 based on various economic and mathematical models, as well as the development of events according to three options: "optimistic", "pessimistic" and "most likely" forecast of various economic indicators. To forecast the "Revenue" indicator for three options (optimistic, pessimistic and most likely), the models of trend analysis and expert assessments were used, taking into account the above correction of the data for 2020. To forecast the indicators "Cost of sales", "General and administrative expenses", as well as "Selling and distribution expenses", structural models were used (relative to the indicator "Revenue") under the assumption that events develop according to one of three possible scenarios (optimistic, pessimistic and most likely) for each variant of the modeled indicator "Revenue". To forecast a number of indicators that are of insignificant importance in the specified form of financial statements, expert assessment methods were used. The results of calculations using the proposed models are presented by nine versions of the forecast "Consolidated statement of comprehensive income". For projections of "Assets", "Equity" and "Liabilities" in "Consolidated statement of financial position", trend, structural and balance models were used in accordance with the data obtained in the forecast "Consolidated statement of comprehensive income".

Keywords: Applied forecasting methodology · Predictive financial reporting · Industrial enterprise · The impact of the COVID-19 factor

© The Author(s), under exclusive license to Springer Nature Switzerland AG 2022
A. Gibadullin (Ed.): DITEM 2021, LNNS 432, pp. 256–268, 2022.
https://doi.org/10.1007/978-3-030-97730-6_23

1   Introduction

Using the actual accounting (financial) statements of industrial enterprises, most stakeholders can perform certain economic calculations for making various management decisions based on the data of these statements. As a rule, these are classic methods of financial analysis that are studied by students, for example, in the course "Business Analysis Using Financial Statements", Massachusetts Institute of Technology (MIT), "MIT Open Courses".[1]) [1, p. 1].

All interested parties, including actual and potential investors, performing various analytical procedures (horizontal, vertical, trend, as well as ratio analysis for the "Consolidated Statement of Financial Position" and for the "Consolidated Statement of Comprehensive Income") have the opportunity to get an idea of not only the actual state of affairs in the studied industrial company, but also about the prospects of its activities in the short term. To do this, you need to follow the procedures for forecasting the accounting (financial) statements for the next financial year. At the same time, the calculated forecast data should closely correspond to the actual values of the key economic indicators of the studied industrial enterprises, which will be achieved in the near future: "Revenue", "Cost of sales", other types of operating and financial income and expenses, various types of profit (gross, operating, before income tax and net), as well as total assets, liabilities and capital.

For several years, the authors have been conducting research on the analysis of the actual reporting of the largest industrial companies both in Russia and the world, and also made calculations of the forecast reporting of the companies under study for the short term, then comparing the calculated results with actual data[2].

In the study of the reporting of industrial companies in Russia, we used data on the activities of all 156 largest industrial companies in Russia (out of 17 industries) included in the rating of 400 largest companies in the country "Expert-400" [3, p. 1] (see Table 1).

It should be emphasized that, according to the Expert-400 rating, the revenue of many industrial companies in 2020 was significantly lower than the same indicators for the previous year[3], and in general, for all 400 largest Russian companies from all industries, revenue decreased by 6.0% [3, p. 1].

The significant reduction in the revenue of many Russian industrial companies in 2020 relative to the previous period was due, first of all, to the restrictions introduced in the economic life of many countries in 2020 due to COVID-19.

---

[1] "MIT OpenCourse" is one of the world's most popular educational Internet resources, with 1.3 billion users browsing the site since its launch in 2002 [2, p. 7].

[2] This analytical work was also attended by students studying the disciplines "Financial Strategy, Planning and Budgeting", "Financial Planning and Forecasting" and "Strategic Financial Management" at the Department of Financial and Investment Management at the "Financial University under the Government of the Russian Federation" during the 2020–2021 and 2021–2022 academic years.

[3] So, for example, in oil and oil and gas companies, revenue in 2020 decreased from 9.7% to 44.6% of the total revenue for 2019, in ferrous metallurgy it fell to 28.2%, in the chemical and petrochemical industry - up to 24.7%.

Table 1. Aggregate economic indicators on the activities of all 156 largest industrial companies in Russia included in the Expert-400 rating for 2020, billion rubles.[a]

No.	Sector of industry[b]	Number of companies	Revenue	Net profit	ROS[c]
1	Light industry	1	40.3	2.0	4.96%
2	Forestry, woodworking and pulp and paper industries	5	313.2	19.8	6.32%
3	Mechanical engineering	30	4 763.9	−134.2	−2.82%
4	Diversified holdings	2	3 137.8	111.2	3.54%
5	Oil and gas industry	19	22 355.4	1 092.0	4.88%
6	Food industry	19	1 648.8	109.3	6.63%
7	Printing industry	1	45.9	7.4	16.12%
8	Manufacture of weapons and ammunition	3	721.7	31.8	4.41%
9	Precious metals and diamonds industry	5	977.4	299.1	30.60%
10	Building materials industry	4	335.0	33.9	10.12%
11	Tobacco industry	5	953.8	48.1	5.04%
12	Coal industry	6	832.2	29.2	3.51%
13	Pharmaceutical industry	1	96.2	16.8	17.46%
14	Chemical and petrochemical industry	16	1 920.4	168.9	8.80%
15	Non-ferrous metallurgy	7	3 446.3	442.4	12.84%
16	Ferrous metallurgy	14	3 891.5	348.5	8.96%
17	Power engineering	18	3 915.8	261.6	6.68%
	Total	156	49 395.6	2 887.8	5.86%

[a] The calculations in Table 1 were made by the authors based on data on the activities of 156 industrial companies from the Expert-400 rating [3, p. 1].
[b] The list of industries in Table 1 is presented in alphabetical order.
[c] ROS – Return on Sales.

In this regard, when forecasting financial statements for 2022, using the initial data in the previous financial statements (for 2021 and 2020), it is advisable to eliminate the influence of the COVID-19 factor on the forecasted revenue of an industrial company.

The specified analytical work allowed the authors to improve the applied methodology for generating predictive financial statements of an industrial enterprise, taking into account the influence of the COVID-19 factor.

The modeling mechanism is presented on the example of actual data of the largest metallurgical company in Russia and the world - PJSC Magnitogorsk Iron & Steel Works and Subsidiaries, since one of the authors works in the said company.

2 Materials and Methods

According to PJSC Magnitogorsk Iron & Steel Works, the largest metallurgical company in Russia and the world, its revenue, for example, for the 3rd quarter of 2020 amounted to only 77.9% and 51.63% of the corresponding indicators for 2019 and 2021 [4, p. 29]. Without a doubt, such a significant impact of the COVID-19 factor requires adjustments to the baseline data for 2020 in order to model the performance parameters for 2022, taking into account the models built on the basis of the trend dynamics of the baseline indicators.

Based on the data of the consolidated interim financial statements of PJSC MMK for the 9 months ended September 30, 2021 and 2020 [5, 6, pp. 6–7], the authors made forecast calculations for 9 months of 2022, taking into account adjustments to the actual reporting due to the influence of the COVID-19 factor in 2020 based on economic and mathematical models, as well as the development of events in three options: "optimistic", "pessimistic" and "most likely" forecast of various economic indicators (see Fig. 1).

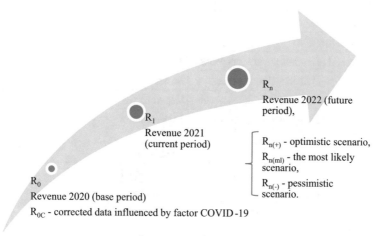

R_n
Revenue 2022 (future period),

R_1
Revenue 2021 (current period)

$R_{n(+)}$ - optimistic scenario,
$R_{n(ml)}$ - the most likely scenario,
$R_{n(-)}$ - pessimistic scenario.

R_0
Revenue 2020 (base period)
R_{0C} - corrected data influenced by factor COVID-19

Fig. 1. Forecasting scheme "Revenue" for 2022, taking into account the impact of the factor COVID-19.

It is advisable to forecast financial statements in the following sequence (see Fig. 2):

1) Forecasting "Consolidated statement of comprehensive income" (see Table 2);
2) Forecasting the "Consolidated statement of financial position" (see Table 3).

To predict the "Revenue" for three options (optimistic, pessimistic and most likely), the authors used a trend analysis model and an additional model of expert assessments taking into account the above-mentioned data correction for 2020 (up to 35% of the actual amount of revenue for 2020) based on the following formulas:

$$R_{GR} = \frac{R_1}{R_{0C}} * 100\% \text{ or} \tag{1}$$

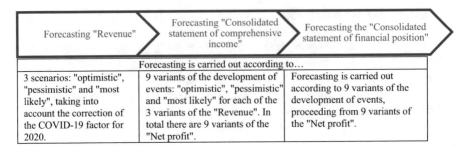

Fig. 2. Forecasting scheme for accounting (financial) statements of an industrial enterprise.

Table 2. The procedure for the forecast of indicators of "Consolidated statement of comprehensive income".

No.	Indicators	Type of economic and mathematical model
1	Revenue	Trend model
2	Cost of sales	Structural model
3	Gross profit	Additive model
4	General and administrative expenses	Structural model
5	Selling and distribution expenses	Structural model
6	Change in expected credit loss, nett	Expert model
7	Other operating income/(expenses), net	Expert model
8	Operating profit	Additive model
9	Finance income	Expert model
10	Finance costs	Expert model
11	Impairment and provision for site restoration	Expert model
12	Foreign exchange loss, net	Expert model
13	Other expenses	Expert model
14	Profit before income tax	Additive model
15	Income tax (20%)	Multiplicative model
16	Profit for the period (net profit)	Additive model

$$R_{GR} = \frac{R_1 - R_{0C}}{R_{0C}} * 100\%, \qquad (2)$$

$$R_{(+)} = R_{GR} \rightarrow max, \qquad (3)$$

$$R_{(-)} = R_{GR} \rightarrow min, \qquad (4)$$

Table 3. The procedure for the forecast of indicators of "Consolidated statement of financial position".

No.	Indicators	Type of economic and mathematical model
1	Total equity	Additive model
2	Detailing of various types of equity	Additive model
3	Total non-current liabilities	Expert model
4	Detailing of various types of non-current liabilities	Additive and structural models
5	Total current liabilities	Expert model
6	Detailing of various types of current liabilities	Additive and structural models
7	Total equity and liabilities	Additive model
8	Total assets	Balance model
9	Total non-current assets	Expert model
10	Detailing of various types of non-current assets	Additive and structural models
11	Total current assets	Expert model
12	Detailing of various types of current assets	Additive and structural models

$$R_{(ml)} = \frac{R_{(+)} + R_{(-)}}{2}, \qquad (5)$$

where R_{GR} – the revenue growth rate; R_1 – revenue for the current period, 2021; R_{OC} – revenue for the previous period, 2020, adjusted for the impact of the factor COVID-19; $R_{(+)}$, $R_{(-)}$ and $R_{(ml)}$ – forecasted revenue for 2022, respectively, determined taking into account the opinion of experts on three options for the development of events ("optimistic scenario" ($R_{(+)}$), "pessimistic scenario" ($R_{(-)}$) and "most likely" ($R_{(ml)}$).

The created group of experts (when the expert forecasting method is used) should be assessed according to the level of competence of experts on the issues under consideration by the formula:

$$K_{C_i} = \frac{K_{A_i} + K_{Aw_i}}{K_{A_{max}} + K_{Aw_{max}}}, \qquad (6)$$

where K_{C_i} – the competence coefficient of the i-th expert; K_{A_i} – argumentation coefficient of the i-th expert; K_{Aw_i} – awareness ratio of the i-th expert; $K_{A_{max}}$, $K_{Aw_{max}}$ – the maximum possible estimates (usually equal to 1) [7, p. 40].

To assess the representativeness of the expert group (M), as a rule, the following formula is used:

$$M = \frac{1}{n} * \sum_{i=1}^{n} K_{C_i}, \qquad (7)$$

$$0.67 \leq M \leq 1, \tag{8}$$

where n – number of experts; at $M < 0.67$ – the expert group is not representative [7, p. 40].

The obtained individual judgments of experts on the issue under consideration are generalized. At the same time, it is important to assess the degree of consistency of expert opinions using the variance (σ^2), standard deviation (σ) and coefficient of variation (ϑ), which are calculated using the following formulas:

$$\sigma^2 = \frac{\sum_{i=1}^{n} (x_i - \overline{X})^2}{n}, \tag{9}$$

$$\sigma = \sqrt{\sigma^2}, \tag{10}$$

$$\vartheta = \frac{\sigma \, 100}{\overline{X}}, \tag{11}$$

where x_i – the assessment of the i-th expert; $\overline{X}$ – the average expert assessment, found by the formula of the arithmetic mean [7, pp. 40–41].

To forecast the indicators "Cost of sales", "General and administrative expenses", as well as "Selling and distribution expenses", structural models were used (relative to the indicator "Revenue") under the assumption that events develop according to one of three possible scenarios (optimistic, pessimistic and most likely) for each variant of the modeled indicator "Revenue" (see Table 4).

Table 4. Matrix of 9 options for modeling forecast accounting (financial) statements of an industrial enterprise.

Consolidated statement of comprehensive income								
Revenue (+) in optimistic scenario $R_{(+)}$			Revenue (ml) in the most likely scenario $R_{(ml)}$			Revenue (−) in pessimistic scenario $R_{(−)}$		
1	2	3	4	5	6	7	8	9
Optimistic scenario	The most likely scenario	Pessimistic scenario	Optimistic scenario	The most likely scenario	Pessimistic scenario	Optimistic scenario	The most likely scenario	Pessimistic scenario
Cost of sales (+) $CS(+)1$	Cost of sales (ml) $CS(ml)2$	Cost of sales (−) $CS(−)3$	Cost of sales (+) $CS_{(+)4}$	Cost of sales (ml) $CS_{(ml)5}$	Cost of sales (−) $CS_{(−)6}$	Cost of sales (+) $CS_{(+)7}$	Cost of sales (ml) $CS_{(ml)8}$	Cost of sales (−) $CS_{(−)9}$
...	...	...	...	...	...	...	...	...
Profit$_1$	Profit$_2$	Profit$_3$	Profit$_4$	Profit$_5$	Profit$_6$	Profit$_7$	Profit$_8$	Profit$_9$
Consolidated statement of financial position								
Equity$_1$	Equity$_2$	Equity$_3$	Equity$_4$	Equity$_5$	Equity$_6$	Equity$_7$	Equity$_8$	Equity$_9$
Liabilities$_1$	Liabilities$_2$	Liabilities$_3$	Liabilities$_1$	Liabilities$_2$	Liabilities$_3$	Liabilities$_1$	Liabilities$_2$	Liabilities$_3$
Assets$_1$	Assets$_2$	Assets$_3$	Assets$_4$	Assets$_5$	Assets$_6$	Assets$_7$	Assets$_8$	Assets$_9$

The structural model of cost of sales is defined as follows:

$$SM_{CS_0} = \frac{CS_o}{R_0} * 100\%, \tag{12}$$

$$SM_{CS_1} = \frac{CS_1}{R_1} * 100\%, \tag{13}$$

$$SM_{CS_{0C}} = \frac{CS_{oC}}{R_{0C}} * 100\%, \tag{14}$$

where SM_{CS} – structural model of cost of sales; CS – cost of sales; R – revenue; index "0" – base period; index "1" – the current period; index "$0C$" – the data of the base period are corrected for the influence of the factor COVID-19.

Variants of the predicted value of the indicator "structural model of cost of sales" are determined taking into account the opinion of experts on three scenarios ("optimistic scenario" $(CS_{(+)})$), "pessimistic scenario" $(CS_{(-)})$ and "most likely" $(CS_{(ml)})$[4]:

$$SM_{CS(+)} = SM_{CS} \rightarrow min, \tag{15}$$

$$SM_{CS(-)} = SM_{CS} \rightarrow max, \tag{16}$$

$$SM_{CS(ml)} = \frac{SM_{CS(+)} + SM_{CS(-)}}{2}. \tag{17}$$

Similarly to formulas 12–17, "structural models" are determined for other indicators of the "Consolidated statement of comprehensive income" if structural models are suitable for forecasting specific indicators.

Similarly to formulas 3–5, various forecast indicators of "income" are modeled, if trend or expert models are applicable to them.

Tables 5(a, b, c) and 6 presents forecast data for key items of the "Consolidated statement of comprehensive income", which were calculated for 9 scenarios.

Based on the results of the forecast "consolidated statement of comprehensive income", work begins on the forecast of the "consolidated statement of financial position" of the company under study.

Table 7 presents the forecast data for the key items of the "Consolidated statement of financial position", which were calculated for 9 scenarios.

---

[4] It is believed that under the "optimistic scenario" of the development of events, costs should be reduced in order to obtain a larger amount of profit, all other things being equal, and under the "pessimistic scenario" of the development of events, the cost structure, on the contrary, grows.

Table 5. Forecast data for calculating "operating profit" for PJSC Magnitogorsk Iron & Steel Works and Subsidiaries for the 9 months ended 30 September 2022, U.S. Dollars million.

a)

Indicators	Scenarios (for expenses)		
	The optimistic No. 1	The pessimistic No. 2	The most likely No. 3
Revenue (the optimistic scenario)	10 165.20	10 165.20	10 165.20
Cost of sales	5 615.26	5 790.10	5 702.68
Gross profit	**4 549.94**	**4 375.10**	**4 462.52**
General and administrative expenses	195.17	203.30	199.24
Selling and distribution expenses	555.02	609.91	582.47
Change in expected credit loss, net	−1	−4	−2.5
Other operating income/(expenses), net	20	−20	0
Operating profit	**3 818.75**	**3 537.89**	**3 678.32**

b)

Indicators	Scenarios (for expenses)		
	The optimistic No. 4	The pessimistic No. 5	The most likely No. 6
Revenue (the optimistic scenario)	10 165.20	10 165.20	10 165.20
Cost of sales	5 615.26	5 790.10	5 702.68
Gross profit	**4 549.94**	**4 375.10**	**4 462.52**
General and administrative expenses	195.17	203.30	199.24
Selling and distribution expenses	555.02	609.91	582.47
Change in expected credit loss, net	−1	−4	−2.5
Other operating income/(expenses), net	20	−20	0
Operating profit	**3 818.75**	**3 537.89**	**3 678.32**

c)

Indicators	Scenarios (for expenses)		
	The optimistic No. 7	The pessimistic No. 8	The most likely No. 9
Revenue (the most likely scenario)	9 402.81	9 402.81	9 402.81
Cost of sales	5 194.11	5 355.84	5 274.98
Gross profit	**4 208.70**	**4 046.97**	**4 127.83**
General and administrative expenses	180.53	188.06	184.30
Selling and distribution expenses	513.39	564.17	538.78
Change in expected credit loss, net	−1	−4	−2.5
Other operating income/(expenses), net	20	−20	0
Operating profit	**3 533.77**	**3 270.74**	**3 402.26**

Table 6. Forecast data for key items of the "Consolidated statement of comprehensive income" for PJSC Magnitogorsk Iron & Steel Works and Subsidiaries for the 9 months ended 30 September 2022, U.S. Dollars million.

No. scenarios	Revenue	Gross profit	Operating profit	Total of financial and other income and expenses	Profit before income tax	Income tax (20%)	Profit for the period
1	10 165.20	4 549.94	3 818.75	−51	3 767.75	753.55	3 014.20
2	10 165.20	4 375.10	3 537.89	−75	3 462.89	692.58	2 770.31
3	10 165.20	4 462.52	3 678.32	−63	3 615.32	723.06	2 892.26
4	8 640.42	3 867.45	3 248.79	−51	3 197.79	639.56	2 558.23
5	8 640.42	3 718.84	3 003.60	−75	2 928.60	585.72	2 342.88
6	8 640.42	3 793.14	3 126.20	−63	3 063.20	612.64	2 450.56
7	9 402.81	4 208.70	3 533.77	−51	3 482.77	696.55	2 786.22
8	9 402.81	4 046.97	3 270.74	−75	3 195.74	639.15	2 556.60
9	9 402.81	4 127.83	3 402.26	−63	3 339.26	667.85	2 671.41

Table 7. Forecast data for key items of the "Consolidated statement of financial position" for PJSC Magnitogorsk Iron & Steel Works and Subsidiaries for the 9 months ended 30 September 2022, U.S. Dollars million.

No. scenarios	Total non-current assets	Total current assets	**Total assets**	Total equity	Total non-current liabilities	Total current liabilities	**Total equity and liabilities**
1	7 090.92	4 727.28	**11 818.20**	9 078.20	1 040	1 700	**11 818.20**
2	6 860.59	4 573.72	**11 434.31**	8 834.31	1 000	1 600	**11 434.31**
3	6 975.75	4 650.50	**11 626.26**	8 956.26	1 020	1 650	**11 626.26**
4	6 817.34	4 544.89	**11 362.23**	8 622.23	1 040	1 700	**11 362.23**
5	6 604.13	4 402.75	**11 006.88**	8 406.88	1 000	1 600	**11 006.88**
6	6 710.73	4 473.82	**11 184.56**	8 514.56	1 020	1 650	**11 184.56**
7	6 954.13	4 636.09	**11 590.22**	8 850.22	1 040	1 700	**11 590.22**
8	6 732.36	4 488.24	**11 220.60**	8 620.60	1 000	1 600	**11 220.60**
9	6 843.24	4 562.16	**11 405.41**	8 735.41	1 020	1 650	**11 405.41**

3   Results

Based on their research, the authors have developed a detailed applied methodology for modeling the forecast consolidated financial statements of a group of companies (industrial enterprise), taking into account the influence of the COVID-19 factor.

In the process of modeling, it is necessary to provide for a scenario approach according to 9 options, within the framework of which timely management decisions can be made.

Deviations between the obtained options can be quite significant - everything will depend on the simulated parameters (within the framework of structural, additive and balance models).

4   Discussion

The regional and local impact of the COVID-19 crisis is highly heterogeneous, with significant implications for crisis management.

Organization for Economic Co-operation and Development (OECD) Creates COVID-19 Territorial Impact Report [8, p. 1]. This paper takes an in-depth look at the territorial impact of the COVID-19 crisis across its different dimensions: health, economic, social and fiscal.

Experts from the European Parliament's Committee on Industry, Research and Energy (ITRE) made a report entitled "Impact of the COVID-19 pandemic on EU industry", which aims to address the following issues:

1) impact of COVID-19 on the EU economy as a whole and across sectors;
2) impact on strategic value chains; and
3) necessary recovery measures to meet the needs of the EU industry [9, p. 3].

According to KPMG experts, "The COVID-19 pandemic continues to impact companies in different ways. Depending on the industry and the economic environment in which a company trades, the 2021 interim reporting period may paint a better or worse picture compared to the last annual financial statements. Either way, COVID-19 continues to affect the recognition and measurement of most companies' assets, liabilities, income and expenses" [10, p. 1].

Japanese experts performed a prediction of the impact of Covid-19 on the sales of listed companies on the basis of the macroeconomic forecast (using a Bayesian Structural Time Series Model), results calculated by Nomura Securities in February and March 2020, before and after the effects of Covid-19 became apparent [11, p. 1].

Deloitte experts suggested using regression models in forecasts of the company's economic activity, taking into account the factor of COVID-19 [12, p. 3].

5   Conclusion

In general, for the calculations, the authors used the classic forecasting models described in the article "How to Choose the Right Forecasting Technique"//Harvard Business Review (Magazine) [13, p. 10].

COVID-19 has had a significant impact on the global economy in general, and individual companies in particular. "The COVID-19 pandemic crisis and its economic effects mean that investors and other stakeholders need high-quality financial information more than ever" [14, p. 1].

In this regard, heads of companies, potential investors, all interested parties should be able to form predictive financial statements of the investigated company in several scenarios, taking into account the COVID-19 factor in order to make the most informed management decisions.

References

1. Business Analysis Using Financial Statements. Massachusetts Institute of technology (MIT). MIT OpenCourseWare. https://ocw.mit.edu/courses/sloan-school-of-management/15-535-business-analysis-using-financial-statements-spring-2003/index.htm. Accessed 26 Nov 2021
2. 2020 OCW Impact Report, MIT OpenCourseWare, pp. 1–12 (2021). https://ocw.mit.edu/about/site-statistics/2020-19_OCW_supporters_impact_report.pdf. Accessed 26 Nov 2021
3. Expert-400: rating of the largest companies in Russia. https://expert.ru/expert/2021/43/spetsdoklad/41/. Accessed 20 Oct 2021
4. Presentation of the financial results of PJSC Magnitogorsk Iron & Steel Works and Subsidiaries for 9 months of 2021, pp. 1–34. https://mmk.ru/ru/investor/results-and-reports/financial-results/. Accessed 25 Nov 2021
5. Public Joint Stock Company Magnitogorsk Iron & Steel Works and Subsidiaries. Unaudited Condensed Consolidated Interim Financial Statements For the Three and Nine Months Ended 30 September 2021, pp. 1–21. https://mmk.ru/ru/investor/results-and-reports/financial-results/. Accessed 25 Nov 2021
6. Public Joint Stock Company Magnitogorsk Iron & Steel Works and Subsidiaries. Unaudited Condensed Consolidated Interim Financial Statements For the Three and Nine Months Ended 30 September 2020, pp. 1–21. https://mmk.ru/ru/investor/results-and-reports/financial-results/. Accessed 25 Nov 2021
7. Kasperovich, S.A.: Forecasting and planning of the economy: a course of lectures for students of specialties "Economics and enterprise management", "Accounting, analysis and audit", "Management", "Marketing". - Minsk: BSTU (Belarusian State Technological University), p. 172 (2007)
8. The Territorial Impact of COVID-19: Managing the Crisis and Recovery across Levels of Government, OECD, 10 May 2021, pp. 140. https://read.oecd-ilibrary.org/view/?ref=1095_1095253-immbk05xb7&title=The-territorial-impact-of-COVID-19-Managing-the-crisis-and-recovery-across-levels-of-government. Accessed 26 Nov 2021
9. Impacts of the COVID-19 pandemic on EU industries. European Parliament's committee on Industry, Research and Energy (ITRE), March 2021, pp. 1–86. https://www.europarl.europa.eu/RegData/etudes/STUD/2021/662903/IPOL_STU(2021)662903_EN.pdf. Accessed 26 Nov 2021
10. Kegalj G. What is the impact of COVID-19 on interim financial statements? https://home.kpmg/xx/en/home/insights/2020/03/covid-19-interim-reporting-10a.html. Accessed 26 Nov 2021
11. Predicting the Industry Impacts of Covid-19 Using a Bayesian Structural Time Series Model. NRI, 2020/04/30, pp. 1–9. https://www.nri.com/en/keyword/proposal/20200430. Accessed 27 Nov 2021

12. Covid-19 Economic Analysis. Deloitte (2021), pp. 1–6. https://www2.deloitte.com/content/dam/Deloitte/global/Documents/Tax/gx-covid-19-economic-analysis.pdf. Accessed 27 Nov 2021

13. Chambers, J.C., Mullick, S.K., and Smith, D.D.: How to Choose the Right Forecasting Technique. Harvard Business Review (Magazine), July 1971. https://hbr.org/1971/07/how-to-choose-the-right-forecasting-technique. Accessed 27 Nov 2021

14. Gould, S., Arnold, C. The Financial Reporting Implications of COVID-19. https://www.ifac.org/knowledge-gateway/supporting-international-standards/discussion/financial-reporting-implications-covid-19. Accessed 28 Nov 2021

Author Index

A. Gibadullin (Ed.): DITEM 2021, LNNS 432, pp. 269–270, 2022.
https://doi.org/10.1007/978-3-030-97730-6